The Birth of the Steam Locomotive – a new history

Star of the film *The Titfield Thunderbolt*, former Liverpool & Manchester Railway No.57 *Lion*, built in 1838, was restored to working order to lead the *Rocket 150* cavalcade in 1980 to commemorate the opening of the line.

Lion subsequently went on tour; here she is on the Keighley & Worth Valley Railway entering Oxenhope station hauling two replica L&MR coaches on 27 August 1981.

Reproduced by kind permission of David C. Rodgers.

The Birth of the Steam Locomotive – a new history

– **Christopher Stead** –

Fern House

First published in 2002 by
Fern House
High Street Haddenham
Cambridgeshire CB6 3XA

A catalogue record for this book
is available from
The British Library

ISBN 1 902702 08 5

Jacket design by Charlotte Edwards
Printed in England by TJ International Ltd

Contents

	List of illustrations	vi
	Introduction	ix
1	Steam power from antiquity to 1750	1
2	James Watt	9
3	Richard Trevithick	16
4	Murray and Blenkinsop	25
5	Some important improvements	31
6	George Stephenson at work	37
7	Locomotion	45
8	Paths of progress	51
9	Rainhill	57
10	Developments in England	67
11	Beginnings in America	72
12	Broad gauge: Valve gears	84
13	The Battle of the Gauges	90
14	Records and regular running	98
15	Conclusion	103
	Bibliography	110
	Index	111
	Acknowledgements	115
	A note on units	115

List of illustrations

	Frontispiece: *Lion*	
1	Hero's Æolipile	2
2	Branca's turbine	3
3	Savery's engine	5
4	Newcomen's engine (1712)	7
5	Watt's double-acting engine (1784) showing parallel motion	9
6	Watt's parallel motion	12
7	Watt's sun and planet 'crank'	12
8	Murdock's road locomotive	13
9	Murdock's road locomotive in section	14
10	Richard Trevithick	16
11	Trevithick's locomotive: the 'Llewellyn drawing' (1803)	17
12	Trevithick's Coalbrookdale locomotive (1803)	19
13	Trevithick's Gateshead locomotive (1805)	21
14	Trevithick's London locomotive *Catch-Me-Who-Can*	23
15	Blenkinsop's two-cylinder rack locomotive (1812)	26
16	Hedley's Puffing Billy	29
17	Hedley's engine as rebuilt	30
18	Leupold's steam valve (1725)	31
19	Sectional drawing of a slide-valve	33
20	George Stephenson (1781-1848)	37
21	Stephenson's second locomotive (February 1815)	40
22	Stephenson's third locomotive with steam springs	41
23	*Locomotion* (1825)	45
24	*Royal George* (1827)	52
25	*Lancashire Witch* (1828)	55
26	*Novelty*	59
27	*Novelty* in section	60

29	*Sanspareil*	62
30	*Rocket* original form	64
31	*Rocket* in section	65
32	*Planet* (1830)	68
33	*Patentee* (1833)	70
34	*Stourbridge Lion*	72
35	Camden & Amboy	74
36	*Tom Thumb*	75
37	*Best Friend* (1831)	76
38	*Atlantic* (1832)	77
39	*De Witt Clinton* (1831)	77
40	*John Bull*	77
41	*Old Ironsides* (1832)	78
42	*Experiment* (1832) later rebuilt with eight wheels as *Brother Jonathan*	79
43	Outside-framed American 4–4–0	80
44	Typical American 4–4–0	81
45	*Pioneer* (1836)	81
46	Robert Stephenson (1803-59)	85
47	Stephenson's link motion	88
48	Howe's link motion	88
49	Chaos at Gloucester	91
50	*Centaur*, similar to *Ixion*	93
51	Stephenson long-boiler locomotive	93
52	Crampton locomotive *Courier* (1847)	96
53	GWR timetable (1844)	100
54	*Bulkeley*, a broad-gauge GWR 4–2–2 locomotive as built from 1847 onwards	103
55	*Jenny Lind* (1847)	103

The Author

Professor Christopher Stead carries on the great tradition of railway-minded clergymen combining enthusiasm with learning.

A student of ancient philosophy and Christian doctrine, he has taught at both Oxford and Cambridge, and was President of the Oxford University Railway Society.

In this book he describes the development of the steam engine from the Ball of Æolus to the work of Stephenson and Gooch, not forgetting the American pioneers and, most excitingly, throws new light on some of the early machines.

Introduction

The book lying open before you provides a new introduction to a subject of extraordinary fascination which has been endlessly discussed. Can one possibly say anything new? I think so. Here, I concentrate on the first part of the story: from antiquity to 1850.

This gives me space to deal more fully with subjects that many authors have treated concisely, often with an eye to readers of experience and knowledge equalling their own. I have considered what uninstructed enthusiasts need to know, rather than assuming that they know a good deal already. I explain not only *what* was done, but *why* it was done. And I can promise that I shall challenge accepted opinion, a task that requires accurate logic and skilled assessment of the evidence. These passages need some thought, but I shall not make them more difficult than they need to be. There are some gossipy interludes to provide a little light relief. My experience as a lecturer and tutor has shown me that there are many ways of presenting one's material; as a wise man once said, as a solid, a liquid or a gas: 'Gas for a lecture, liquid for a book, solid for an archangel in retreat'. This book is a fairly concentrated liquid – a cordial, perhaps.

We pass, then, to record an advance; one of the most important advances in the history of human progress.

1 – Steam power from antiquity to 1750

On Tuesday, 21 February 1804, the Cornish engineer Richard Trevithick won a bet of £500. His newly constructed steam locomotive succeeded in hauling a load of 10 tons of iron along the tramway from Penydaren ironworks in South Wales to the canal basin at Abercynon, a distance of nearly 10 miles. This was the first occasion on which a useful load was transported by steam; the point at which the eventful history of railway locomotion begins.

In fact the bet was won by a handsome margin. On the opening run some seventy passengers hitched a lift, adding several tons to the required load. A month later, 25 tons were successfully hauled. But as is generally known the engine's weight of five tons was too much for the lightly-constructed track; it was taken out of locomotive service and used to work a hammer.

Such an invention could not have been the work of a few months. Usually, many years must elapse between the first conception of a new machine and the working out of a practical design and of the means to produce it. The railway locomotive resulted from two lines of descent: the development of the railway track and the application of steam to set the wheels in motion. The first of these need not concern us here; after all, vehicles can be moved along rails by men or horses, by water power, using cables, or even by wind using sails – not to mention the diesel and electric locomotives to which steam has given way. But something must be said about the earlier history of steam as a motive power.

This takes us back to the ancient Greeks, and to Hero of Alexandria, probably first century AD. Hero produced a device called an 'Æolipile' or 'Ball of Æolus' named after the god of the winds (this seems more likely than 'Æolipyle' – 'gate of Æolus' – which some scholars prefer). It was based on the common Greek *lebés*, or cauldron, a round vessel tapering outwards, covered by a lid and resting on a tripod so that a fire could be applied underneath. Hero's device of course called for the lid of the cauldron to be firmly sealed, so that a pressure of steam could develop.

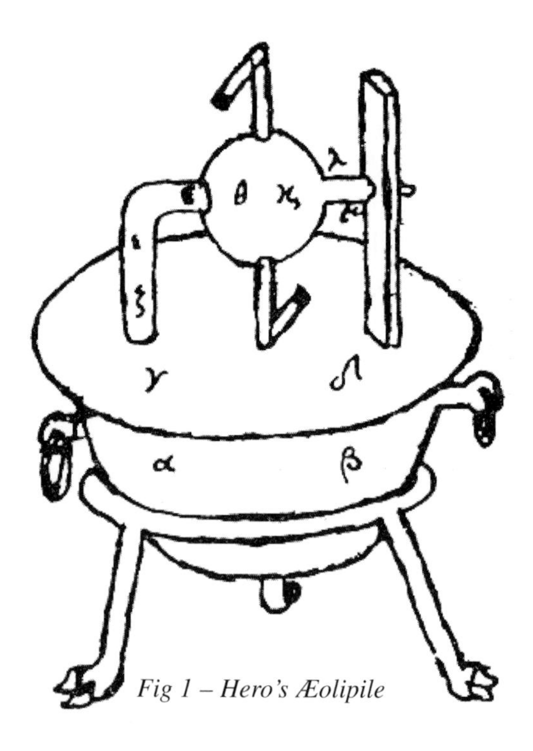

Fig 1 – Hero's Æolipile

From the lid projected two supports, one of which was hollow and conveyed the steam to a hollow ball spinning upon a horizontal axis. To the ball were fixed two L-shaped pipes with open ends turned over to face in opposite directions; the steam issuing from these caused the ball to revolve by reaction, on much the same principle as a rocket is driven upwards.

The *lebés* which functioned as a boiler might indeed have been already provided with a small hole like a modern kettle so that excess steam could escape. In this case it would have been easy to feel the force of the steam-jet. Hero's device would then be a striking anticipation of the story commonly told about James Watt, whose thoughts on the use of steam were prompted, it is said, when he saw the lid of an ordinary kettle agitated by the pressure of the steam within.

Hero's engine was not developed and exploited, since the ancient world had little need for new sources of power. Slave labour was abundant and men and horses could do most of what was required, supplemented by wind or water mills for heavy work such as grinding corn.

But after the upheavals of the middle ages, the recovery of ancient learning brought a new spirit of enquiry and experiment, which extended to the science of mechanics; a striking example is Leonardo's primitive attempt to design a flying machine. A new form of steam turbine was suggested by Giovanni Branca in 1629; this used a jet of steam issuing from a boiler to turn a wheel equipped with vanes like that of a water-mill.

Fig 2 – Branca's turbine

The device itself was much less interesting than Hero's, but a means of use was suggested. The motion was transmitted by reduction gearing to a roller which activated a pair of stamps, apparently for the purpose of pounding drugs in mortars. Yet it can hardly have been effective; even with a reduction gearing of about 64:1, as shown in the drawing, it could not have provided any considerable power.

Other investigators were at work on the fundamental properties of matter, including fire, air and water. In 1606 Giambattista della Porta suggested a practicable method of raising water by steam power. Steam from a boiler was piped into the upper part of a closed vessel containing water; the pressure of the steam forced the water to escape through a second pipe opening below the water level and leading upwards. Another means of raising water was suggested by a second experiment; a flask is filled with steam and plunged neck-downwards into a vessel of water; the steam then condenses, forming a partial vacuum and atmospheric pressure forces the water up into the flask. Della Porta did not attempt to develop this device; the water could not escape from the flask. Nevertheless, he had suggested two totally different methods, both of which were put to practical use by subsequent engineers; but without, as yet, employing any moving parts.

The need for raising water was felt most acutely by mining engineers. The early miners naturally exploited easily accessible coal seams and veins of ore; as these became exhausted it became necessary to dig deeper and the inrush of water in increasing quantities and at greater pressures became a steadily growing nuisance and danger. Pumps were installed, worked by men or by horses; occasionally a water wheel above ground could be used to transmit power to work underground pumps, as did the famous Big Wheel at Laxey in the Isle of Man. When such devices were no longer sufficient, the possibilities of using steam had to be seriously considered, and pumping engines were first introduced in England late in the 17th Century.

Some credit for priority should probably be given to Edward Somerset, Marquis of Worcester (1601–67) who at least claimed to have constructed a machine to raise water by means of steam pressure to a height of 40 feet. This was seen at work at Vauxhall in 1664.

But it was Thomas Savery (?1650–1715) who was the first to manufacture and sell steam pumping engines in considerable numbers. A

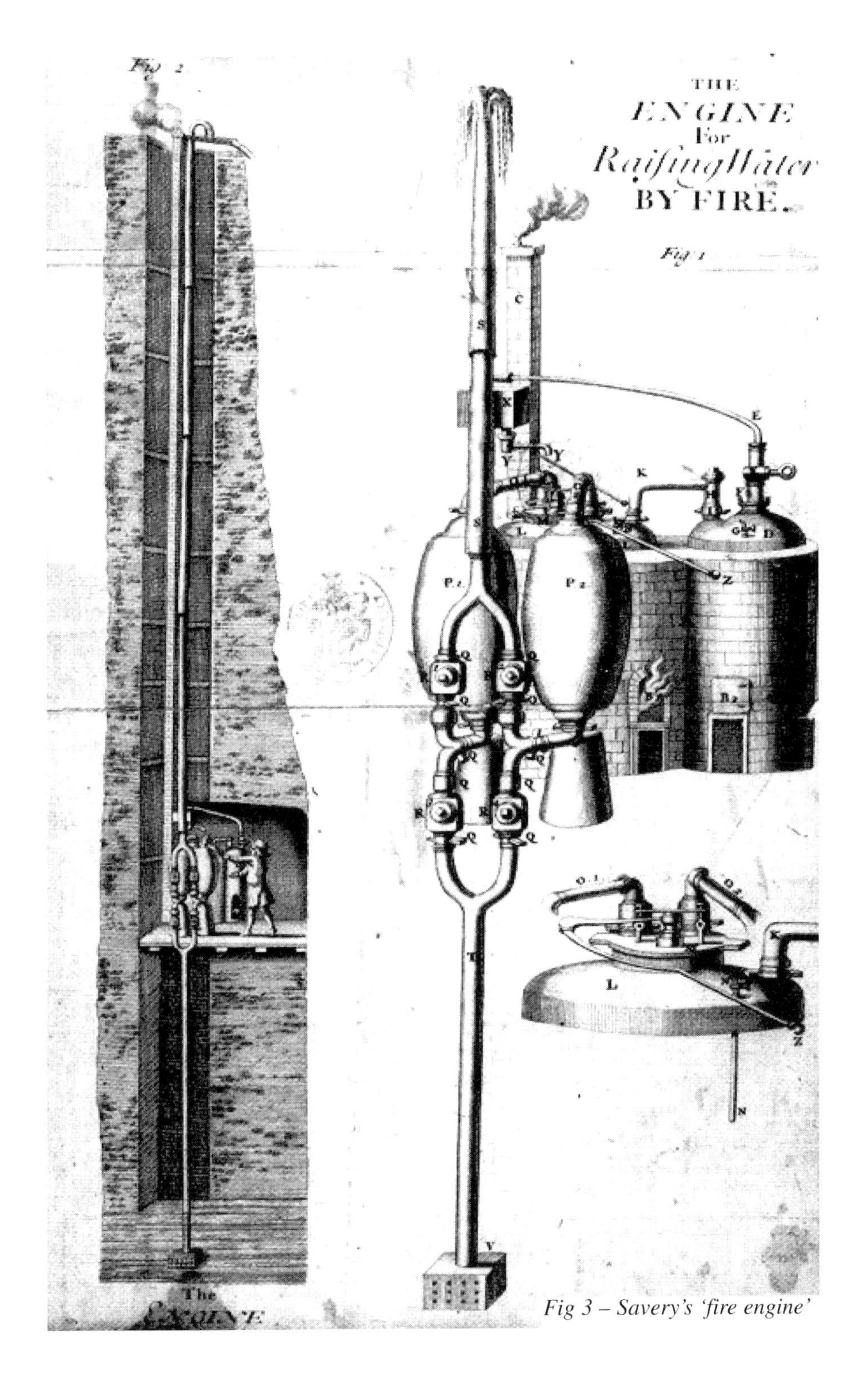

Fig 3 – Savery's 'fire engine'

model of Savery's engine was shown to the Royal Society in 1699, and full-sized engines were being advertised and manufactured in 1702.

Savery's engine combined the two methods suggested by della Porta. Steam from a boiler was led into a receiver; it was then condensed to form a partial vacuum and draw water up by suction. The next step was to admit steam at high pressure and force the water upwards; the total lift was estimated as 16 feet by suction plus 42 feet by pressure, an advance on the original claim. Savery soon introduced improvements for ease of working, and a valuable step was to use two receivers (as illustrated) working alternately, so that water could be delivered in a continuous stream.

Despite Savery's ingenuity, his machine suffered from three incurable defects. First, the height to which water could be drawn by suction was limited, as it depended on atmospheric pressure; trials with a common pump had shown that water would not rise in the suction pipe beyond 30 feet at the most, after which it came to a stand, leaving a vacuum above which quickly became filled with water vapour, allowing the water to fall again. Second, there were insuperable difficulties in constructing a receiver, and still more in the connecting pipes, which could withstand high-pressure steam; the problem was eventually solved but it took many years of effort and experiment. Third, the machine was inherently inefficient, as the admission of high-pressure steam would warm up the receiver and interfere with the cooling process that had to follow in order to condense the steam. The engine could be made to work, but only very slowly.

Meanwhile, a totally different method of raising water was being devised by Savery's contemporary Thomas Newcomen (1663–1729). His machine worked entirely by condensation of steam, using a piston in a cylinder, and so leading by a direct line of descent to the work of Watt, Trevithick and Stephenson. In its essentials, it was an extremely simple device. The vertical cylinder was open at the top, with a piston connected by a rod or chain to one end of a rocking beam. Steam was admitted to the cylinder below the piston at the bottom of its stroke; this caused the piston to rise, aided by the weight of the descending pump-rod hung from the other end of the beam. At the end of the stroke a jet of cold water was squirted into the cylinder, causing the steam to condense, forming a partial vacuum, so that the piston was driven sharply downwards by atmospheric pressure and the pump-rod was raised to draw up the water.

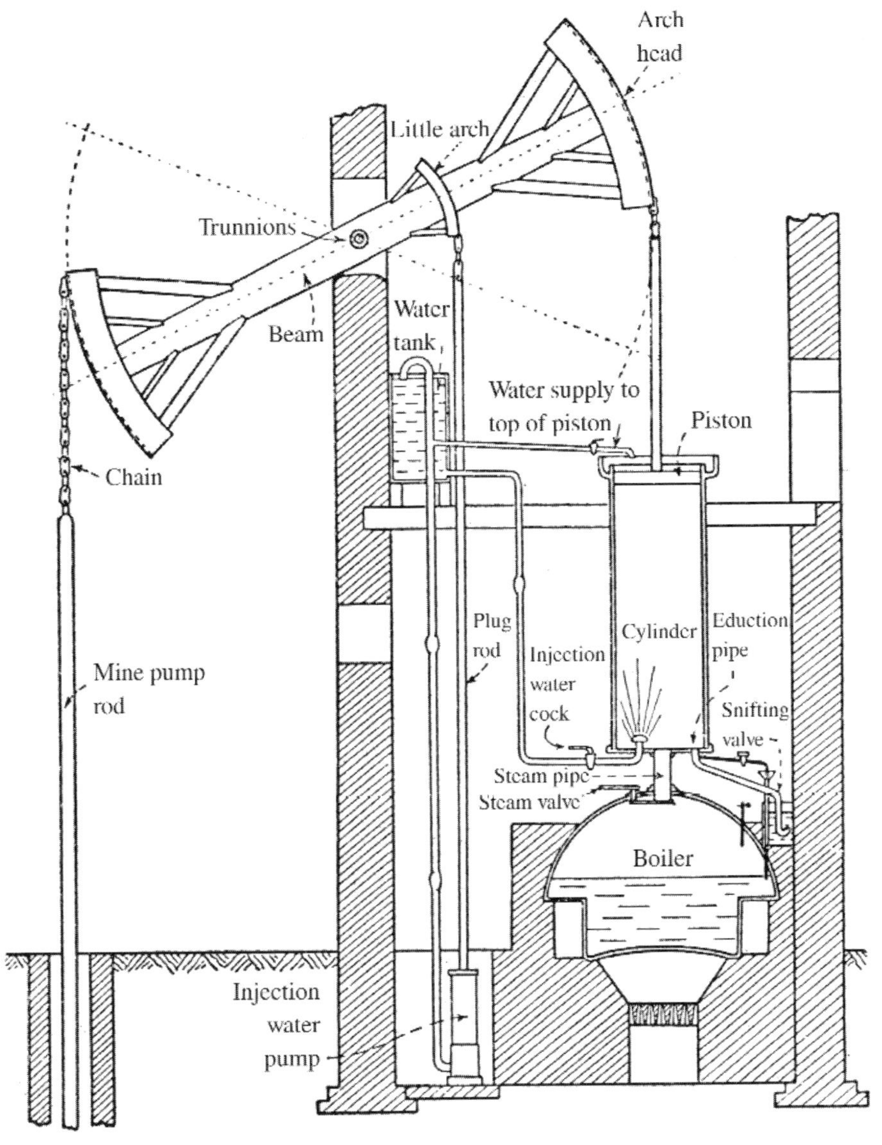

Fig 4 – Newcomen's engine (1712)

Newcomen's engine had several great advantages over its pre-decessors. First, it used low-pressure steam, thus enormously simplifying the task of constructing a suitable boiler and its pipe-work. Second, its power could be increased at will by using a very large cylinder: cylinders seven feet in diameter and eight feet tall were by no means unknown. Third, by using a really long pump-rod, it was now possible to bring water up from a considerable depth, with the engine conveniently located at ground level. Several improvements were soon introduced, some for convenience of working – such as automatic valves; others to deal with inconveniences, such as the snifting valve, designed to get rid of unwanted air which had crept into the cylinder with the steam and could not of course be condensed. In its developed form Newcomen's engine gave useful service for most of a century after its introduction in 1712; it was especially useful to the miners of Cornwall, who were already working at considerable depths and needed an efficient machine in view of the high cost of coal, which could not be found locally and had to be brought in by sea.

For some 60 years, Newcomen's engine had no serious rivals, though its limitations became obvious as the demands on it increased with the growing depth of the mines. It was a pumping engine pure and simple, though wheels could be turned by pumping up water and using it to work a water-wheel. Its defects were first identified, and a remedy suggested, by the brilliant scientific work of James Watt, which led to revolutionary new developments in the construction of steam-engines.

2 – James Watt

Stationary engines developed; locomotives rejected

James Watt was born in 1736 at Greenock on the River Clyde, west of Glasgow. His grandfather Thomas Watt was a mathematician who made his living as a 'teacher of navigation', a subject which was in growing demand as the Clyde was gradually made usable by large merchant ships. His father, the elder James, was described as a 'wright', a constructor who made all sorts of equipment in wood and metal for the fitting-out of sailing vessels; he soon broadened his interests to become a ship-chandler, and later a merchant. He married Agnes Muirhead, a well-conducted and intelligent woman, and the great James Watt was their eldest surviving child.

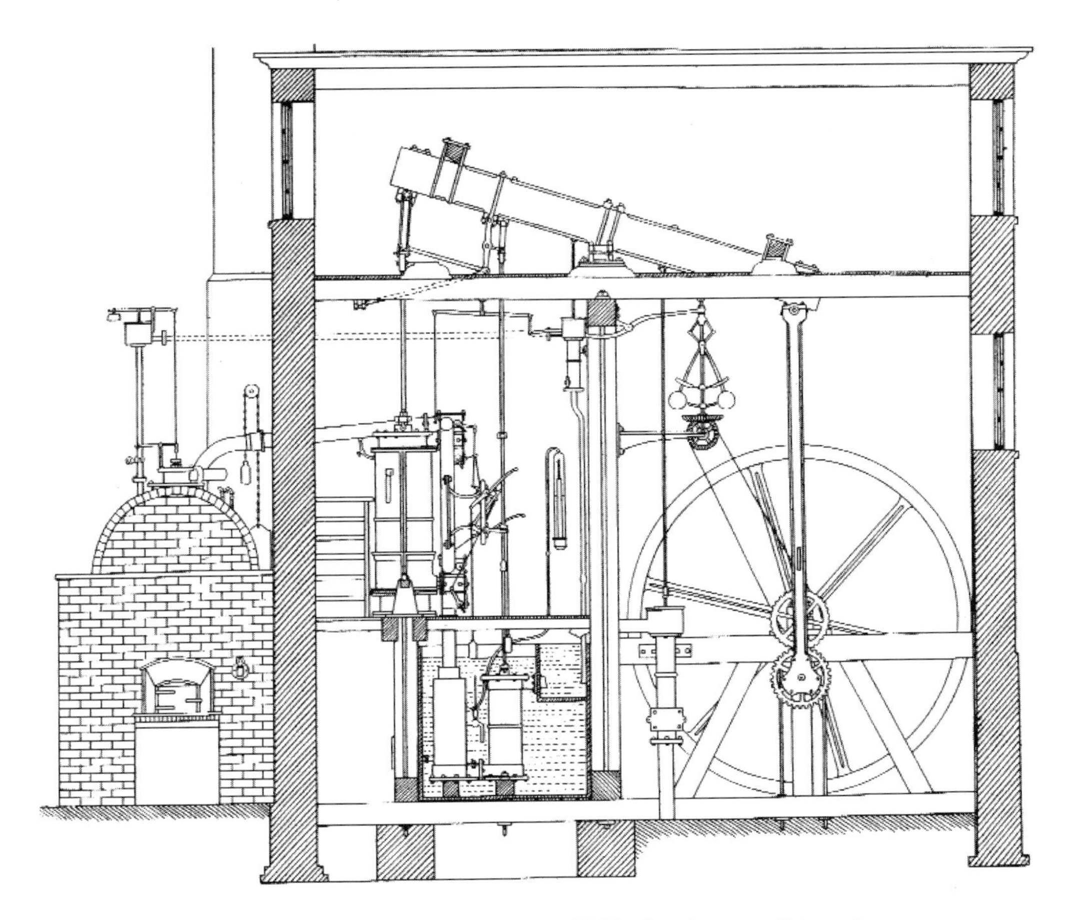

Fig 5 – Watt's double-acting rotative engine (1787) showing parallel motion

Frail in physique and nervous by temperament, he made a poor showing in his early years at school, but at the age of 13 he began to be noticed as an unusually promising mathematician. He also developed extraordinary skill and industry as a craftsman, making all kinds of mechanical models in his home workshop and studying the scientific principles which governed their behaviour. He decided to earn his living as a mathematical instrument maker, but could not acquire the necessary training in Glasgow, and was advised to travel to London. Here he ran into difficulties, largely brought about by the restrictive rules imposed by the trade guilds; but eventually he was taken on by John Morgan at his shop off Cornhill; of whom he wrote: 'though he works chiefly in the brass way, yet he can teach me most branches of the business, such as rules, scales, quadrants, etc.' Watt set to work with prodigious industry, and within the year allotted for his stay in London had completed what was normally three to four years' work. He worked long hours, and rarely went out, partly for fear of being captured by the press gang. But he returned to Glasgow at the age of twenty-one confident that he was now capable of setting up in business.

Here again it was difficult to find an opening, as he had served no apprenticeship and the rules of the trade guilds forbade him to work as a craftsman. But by a great stroke of luck he was taken up by Professor Robert Dick at the University and put to work repairing some astronomical instruments that had been damaged during their voyage from Jamaica. For this purpose he was given the use of a small room in the College where he would be safe from the interference of the guildsmen. Before very long he was making the acquaintance of eminent men of science, and was able to style himself 'Mathematical instrument maker to the University.'

Watt's interests soon diversified; he began to construct musical instruments and devised a perspective drawing machine; in everything he touched he could not be content until he had thoroughly mastered the scientific principles on which it was based. His revolutionary work on the steam-engine began in 1763–4, when he was asked to repair a brass model of a Newcomen engine which failed to work after the first few strokes. Watt's own comment was: 'I set about repairing the model as a mere mechanician', but with characteristic energy he had soon mastered the physical principles which governed the behaviour of water and water-vapour at various temperatures and pressures.

He found that, besides incidental defects, the model showed up an inherent fault in the Newcomen engine when constructed full size. He reasoned that the cylinder needed to be kept hot, in order to avoid premature condensation of the steam; but the process of condensation required the cylinder and its contents to be cold. As a result, the engine was inefficient; heat was wasted by the cylinder being alternately warmed and cooled, and in each case failing to reach the temperature required for efficient working.

The solution occurred to him in May 1765 while he was walking on the College Green. His account of it, as later recorded by a friend, has often been reprinted, but is still worth recalling: 'It was *in the Green of Glasgow*. I had gone to take a walk on a fine Sabbath afternoon ... I was thinking upon the engine at the time, and had gone as far as the Herd's house when the idea came into my mind, that as steam was an elastic body it would rush into a vacuum, and if a communication was made between the cylinder and an exhausted vessel, it would rush into it, and might there be condensed without cooling the cylinder ... I had not walked further than the golf-house when the whole thing was arranged in my mind.'

To test his ideas, Watt immediately set about making a model, which still survives at the Science Museum, and followed this up with a miniature working engine with a cylinder two inches in diameter. He began making drawings for a full-size engine to be erected at Kinneil, using a cylinder with a closed top to retain the heat, the piston passing through a stuffing-box; a later improvement was to surround the cylinder with a steam jacket. But from now on his work was beset by problems and delays. He needed a steady income, whereas his plans for a new engine could not bring in any immediate return. He found work as a surveyor, planning projected canals; this occupied him from 1766 to 1779, with only brief intervals for work on the engine; and his new ideas involved a frustrating search for new materials and difficulties with poor workmanship.

His prospects improved in 1775, when he entered into partnership with the industrialist Matthew Boulton, who foresaw the enormous possibilities of his work and agreed to provide for his needs. It was not till then that the Kinneil engine was eventually freed from its problems, and Watt immediately started work on two larger machines, one with a 50-inch cylinder.

From now on his ideas rapidly bore fruit. An important improvement was the double-acting engine, with steam admitted alternately above and below the piston; its plan is shown on a drawing made for a patent in 1775 (Fig 5), and construction soon followed. From this Watt went on to develop a 'rotative engine' which would turn the wheels of machinery without using the cumbrous expedient of pumps and water-wheels.

But a problem arose at once. The piston must act on the rocking beam; but any point on the beam necessarily moved in an arc, whose centre was the pivot or trunnion of the beam; whereas the piston rod (F in Fig 6) must move straight up and down. Watt's solution was to correct the curvature of the arc by introducing another arc of opposite curvature so that the two motions would cancel out. He used a parallelogram of jointed rods (corners BGDE), whose top corners (BG) were attached to the beam, and whose lower corners (DE) of the parallelogram were guided by a rod (CD) swinging from a fixed pivot (C) in the opposite direction from the trunnion. The parallelogram flexed as the machine went to work, and the piston rod (F), attached to its lower corner (E), was guided in the required vertical path. In practice the rods were often duplicated, so that the pairs embraced each other at the pivots, thus avoiding any tendency for them to twist sideways.

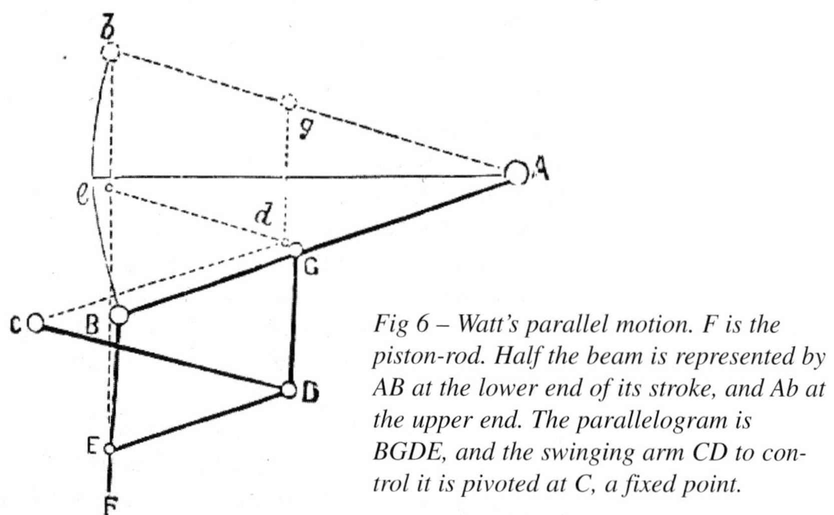

Fig 6 – Watt's parallel motion. F is the piston-rod. Half the beam is represented by AB at the lower end of its stroke, and Ab at the upper end. The parallelogram is BGDE, and the swinging arm CD to control it is pivoted at C, a fixed point.

Watt was rightly proud of this invention, and it had a long life; in a modified form it was used on Stephenson's *Locomotion* of 1825 and Hackworth's *Royal George* of 1826, and it remained in use on the *Puffing Billies* until about 1862. These mechanisms often dispensed with the parallelogram of rods; but the term 'parallel motion' is convenient, and is used for any device which guides the pistons by means of two opposing arcs.

Fig 7 – Sun and planet gear

But a difficulty soon arose with the rotative engine. By some sharp practice on a rival's part, and some misunderstanding on Watt's, he was convinced that a patent existed which forbade him to use the common device of a crank, at least on locomotives; whereas the crank was in regular use on lathes and grinding wheels. Watt soon devised five alternative methods of converting reciprocating motion into rotation; the one he usually adopted was the 'sun and planet'.

A gear-wheel fixed to the connecting rod engaged with a second gear-wheel on the flywheel shaft (Fig 7). This arrangement behaved exactly like a crank, but the supposed requirements of the patent were satisfied.

The rotative engine was soon in great demand, and brought in handsome profits, whereas the pumping engines sent to Cornwall had caused difficulties with the mine-owners, who were troubled by their own problems over the increasing cost of fuel, and were reluctant to pay the licence fees expected of them; the original agreement had given them the option of purchasing an engine outright, but they seldom did so.

Nevertheless, with his conservative temperament, Watt gradually failed to keep up with the needs of the time. He remained firmly attached to the low-pressure engine, with its massive cylinder and bulky condenser; whereas by the 1780s there was vigorous discussion of the possibilities of a locomotive; indeed Watt himself had discussed them with John Robison (Professor of Mechanical Philosophy at Edinburgh) as early as 1759. But it soon became clear that Watt's enormously bulky low-pressure condensing engine was quite unsuitable; it would be essential to use the much smaller and more compact high-pressure engine, which the cautious Watt considered unsafe and refused to adopt.

The first locomotive of this kind was constructed in France by Nicholas Cugnot in 1763. It was designed as a tricycle driven by the front or steering wheel; thus the boiler and cylinder had to be carried on the front axle and swivel round with it, making the whole machine absurdly front-heavy. It went, but proved almost uncontrollable and was soon abandoned. A more

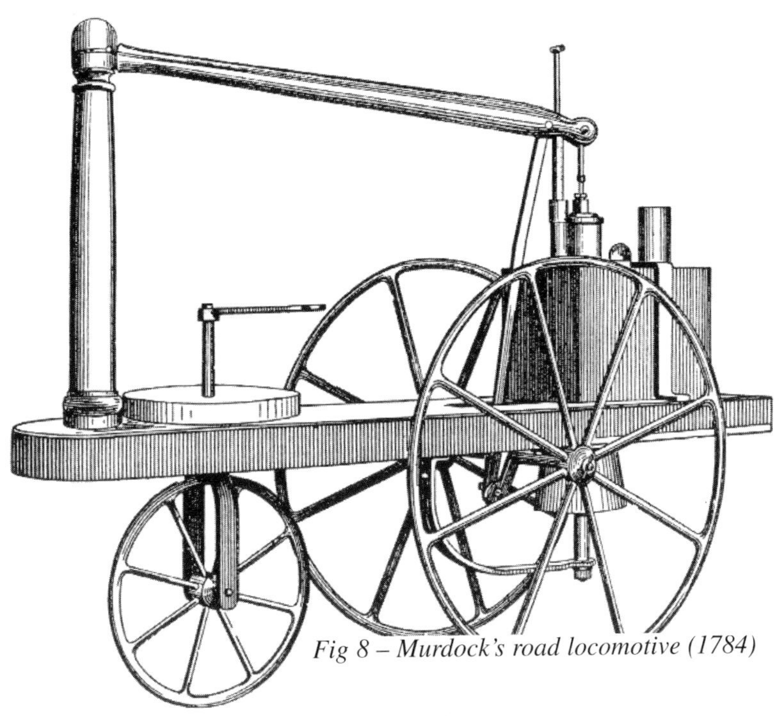

Fig 8 – Murdock's road locomotive (1784)

practical concept was tried out by Watt's gifted assistant William Murdock (1754–1839), who constructed a model road locomotive in 1784. This deserves a fairly detailed description, as it clearly demonstrates Murdock's inventive skill, which is too often passed over.

Like Cugnot, Murdock built a three-wheeler, in which all the wheels would stay on the ground; but the boiler and cylinder were carried on the rear pair of wheels. The front wheel supported a post on which was pivoted the 'grass-hopper beam', in effect one half of the normal two-ended beam; its weight thus ensured sufficient stability for the front wheel; while a connecting rod descending from its rear end turned a crank on the rear axle. The rod passed through a space cut in the solid wooden platform which served as a frame and supported the boiler and cylinder. The fire-tube passed obliquely through the boiler to make room for the cylinder, 0.75 inches in diameter x 2 inches stroke, which was set into it for about half its length. A cylindrical fuel tank was placed below the platform, supported by a double-angled metal rod to leave a clear space below the boiler so that air could mix with the vapour issuing from the tank, and the resulting flame pass through the platform to the fire tube. The piston drove upwards, the weight of the beam assisting the return stroke, as no flywheel was provided. Steam was admitted to the cylinder by a simple form of piston valve driven by a slotted rod running parallel to the connecting rod. The diagram does not make it clear how the exhaust

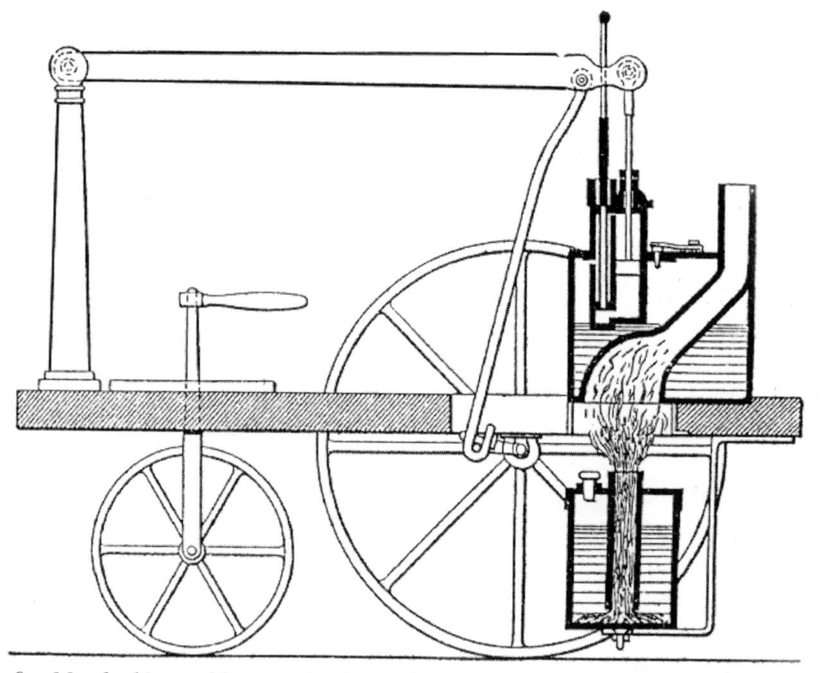

Fig 9 – Murdock's road locomotive in section

steam was allowed to escape.

The machine was tried out on a country lane at night, and the sight of the 'fiery monster' caused great alarm. Watt however was displeased and forbade Murdock to continue such experiments; Murdock loyally obeyed his instructions.

Watt's partnership with Boulton terminated in 1800. In 1790 he had had a comfortable house built near Soho, the site of Boulton's factory; to this he added an estate at Doldowlod in Wales to use as a summer retreat. Boulton died in 1809, while Watt lived on till 1819. Meanwhile his remarkable achievements received ample recognition; he was elected a Fellow of the Royal Society of Edinburgh in 1784 and, with Boulton, of the London Society the next year. He was given the degree of Doctor of Laws, *honoris causa*, by Glasgow University in 1806, and in 1814 became one of only eight men dignified as a Foreign Associate of the French Academy. But by this time the development of the steam engine had passed into other hands.

3 – Richard Trevithick

Richard Trevithick was in many ways the exact opposite of James Watt; a giant of a man, endowed with enormous physical strength; enterprising where Watt was diffident; impulsive where Watt was methodical. His mechanical inventiveness was extraordinary, but it was largely intuitive; he lacked Watt's analytical approach and thorough understanding of the physical principles involved, and although he embarked on a great variety of extremely useful projects, he often lacked the patience to bring them to per-

Fig 10 – Richard Trevithick (1771–1833)

fection. A major exception, perhaps, was his work on a Thames tunnel, which occupied him from 1805 until 1808; though in the end it had to be left unfinished through no fault of Trevithick's.

Trevithick, I think, has been unfairly judged; subsequent writers have been unduly influenced by an obituary published in 1833, by which time the development of the steam locomotive had caught people's attention, and was almost regarded as the test of a successful engineer. Trevithick's work on the locomotive had come to a close by 1808, largely because the essential requirement for its use, a robustly constructed track, was not yet available. And his departure for Peru in 1816 removed him from the English scene at the very moment when a fresh approach to locomotive design was being made by Blenkinsop, Hedley and Stephenson, aided by the much better track introduced by Stephenson and Losh after 1815. The failure of his South American projects was remembered against him, and it was forgotten that his work with locomotives had been something of a side-issue; his principal occupation was the design and manufacture of large numbers of stationary steam engines, which worked efficiently and were in great demand. He returned to England to die in poverty in 1833, at the age of 62, and his reputation suffered the usual fate of the unfortunate. Nothing could be more malicious or more stupid; in actual fact his inventive genius was active to the last.

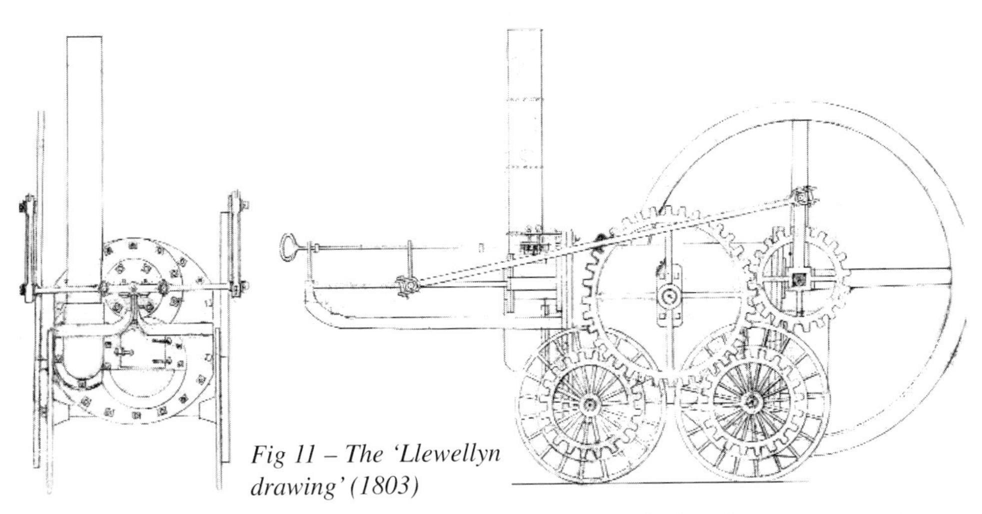

Fig 11 – The 'Llewellyn drawing' (1803)

In February 1831 he took out patents for 'apparatus for heating apartments by warm air; a superheater; jet water-propulsion of vessels; and a boiler and superheater applied to locomotives.' If this is failure, who needs success?

Trevithick had realised from the start that high-pressure steam must be used in order to build an engine that was reasonably compact and light in weight, whether for use as a locomotive or for a portable stationary engine.

He was born in 1771; his interest in locomotives developed quite early, and he built three models to test his ideas, two of which still survive. Both of them show some original features. The second model, of 1796–7, had a single cylinder partly contained within the boiler; it drove upwards on to a long cross-head, from which connecting rods came down both sides of the boiler to drive the wheels and also a flywheel through a system of gearing. It was designed to be fired by a red-hot iron bar placed in a tube within the boiler, so that no flue or chimney was needed. An interesting feature is the provision of a set of jacks, so that the engine could be adapted for stationary use.

The third model, of 1797, was very different; it was planned as a road tractor, with its front wheels mounted on a swivel for steering. The cylinder and cross-head were similar, but the drive was taken directly to crank-pins on the rear wheels, and no flywheel was provided. The boiler once again was cylindrical in form, placed horizontally, but with its lower half partitioned off to form a fireplace, from which the hot gases escaped through holes in the side. The fire could thus be refuelled, but it lacked the advantage of being wholly enclosed within the water space.

Trevithick then embarked on a number of full-size machines, beginning with two road carriages in 1801–3, followed by four railway locomotives in 1803–8. After many months of preparation, the first road carriage was tried out at Camborne in Cornwall. It had a brief and adventurous career. On its trial trip it ran out of steam. On the next run it overturned on coming to a gully in the road which jerked the steering handle out of the driver's hands. It was righted, and placed in a shed adjoining an inn, to which the party then adjourned for an ample dinner of roast goose. The fire meanwhile remained alight and the water all boiled away, causing a conflagration which consumed both the shed and the engine itself.

In contrast, the second carriage was taken to London, where it ran quite successfully for some months, though without attracting much public interest. We know something about its construction, since it almost certainly agreed with a carefully-executed drawing prepared for a patent specification in 1802. This shows a fairly small boiler of the normal 18th-century type, with the fire underneath, but equipped with a casing which forced the hot gases to pass up the sides of the boiler on their way to the chimney. This arrangement, again, had been tried before. There was a single horizontal cylinder with the drive taken to a crank on the rear axle, to which the wheels were rigidly attached. These driving wheels were made very large, so that they could easily pass over pot-holes in the road. An appended drawing shows a more sophisticated type of boiler, with the hot gases conducted by a flue through the water space; but it does not suggest that this was used on a road carriage.

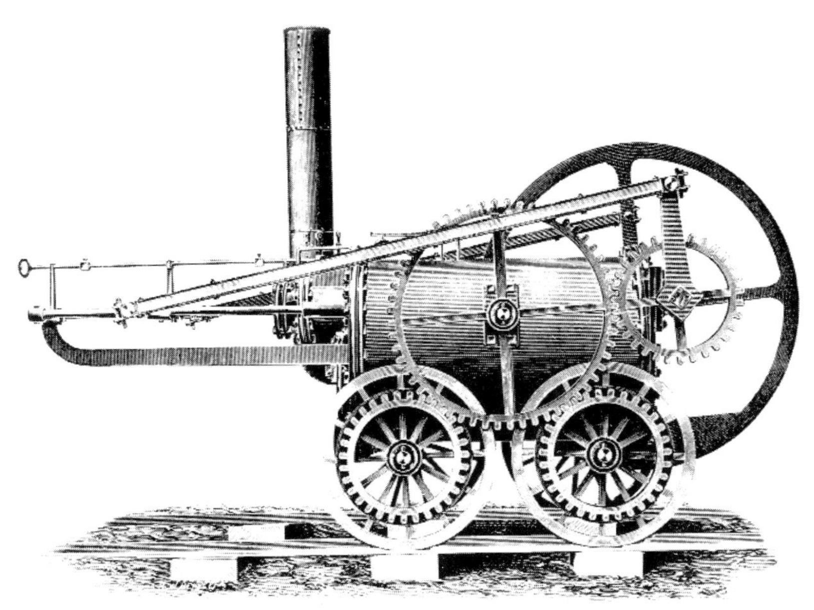

Fig 12 – Coalbrookdale locomotive (1803) after Llewellyn; often wrongly supposed to be the Penydaren locomotive (1804)

In contrast, the construction of the Camborne road carriage is almost entirely unknown. A conjectural drawing exists, made many years later by Francis Trevithick and relying on the memories of old men who had seen it. But I do not think it can be trusted. Its proportions are ungainly, with a large and necessarily heavy boiler carried on improbably small wheels. Such a machine, I consider, would have been too unstable to survive even a first trial run over the uneven roads available. Moreover it shows a return-flue boiler, which we know was used on the Coalbrookdale locomotive of 1803, as will soon be described. But it is hard to see why this improved type of boiler should have been used in 1801, discarded in 1802 and reintroduced in the three railway locomotives of 1803–6. More probably, I think, the old men's memories were influenced by their familiarity with these locomotives.

Our knowledge of the first three depends partly on some details relating to the Penydaren locomotive given by Trevithick in letters to correspondents, and partly on two drawings. One of these is a carefully-executed contemporary drawing showing an engine with flanged wheels running on edge rails; this must be a third locomotive, built at Gateshead in 1805. The two others are known to have run on plate-ways: these are: first, the long-disregarded engine built for a 3-foot gauge plate-way at Coalbrookdale in 1803; and second the much larger Penydaren engine of 1804, which ran on a 4ft 2in plate-way. The second drawing, known as the 'Llewellyn drawing', is clearly a copy made from an earlier original, and so is rather less reliable, though generally accurate. It has often been taken to represent the Penydaren engine, and has been often reproduced with this description. But research has shown that this identification is impossible, as we shall explain. Before doing so, it will be convenient to set out the main features indicated in both drawings.

The machinery shown has a new layout. There is a single cylinder, mounted horizontally and partly enclosed within the boiler, partly projecting at one end. It has Trevithick's familiar arrangement of very long cross-heads driving a connecting-rod on each side of the boiler. One of these works with a crank-pin on the flywheel: the other leads to a crank connected by gearing to the driving wheels. A very curious feature of the Coalbrookdale engine is that these wheels revolve loosely on rigid axles, so that the drive is taken to two wheels on one side of the engine only, although the patent drawing of 1802 had called for wheels rigidly fixed to a revolving axle. The one-sided drive is clearly shown on a model made for the Science Museum and misleadingly identified with the Penydaren engine; it is perhaps explained by seeing the earlier machine as having been rather hastily improvised on the basis of Trevithick's more widely-used stationary engines. The Gateshead locomotive certainly did not have the crude farm-cart arrangement of a cotter-pin slotted into the axle; the drawing shows what is apparently a square-headed bolt, presumably making a rigid connection of a revolving axle with its wheels. I can find no evidence to show what arrangement was used on the Penydaren engine

The boiler, however, was efficient, as well as being simply constructed and maintained, and was very likely Trevithick's own invention. In an improved form it is best seen on a stationary engine of 1805; a simple cast-iron pot, U-shaped in cross-section and flanged at its open end. To this was bolted a cover-plate on which were mounted both the cylinder and the internal flue; this was a U-shaped tube of wrought iron, enlarged at one end to contain the fireplace, and narrowing as it returned towards the chimney-base, which was placed immediately beside the fire-door. This must have been a rather cramped arrangement in the small-sized boiler used at Coalbrookdale, but it had the great advantage that the entire contents of the boiler could be withdrawn for inspection simply by unbolting the cover-plate, while the furnace and flue are entirely contained within the water space. The exhaust steam was turned into the chimney to draw the fire.

The locomotive boilers were differently constructed. The Coalbrookdale one was cylindrical, probably made of wrought iron, and flanged at both ends to take flat cover-plates, with the working cylinder embedded in one of them. This arrangement can be seen on the replica recently made for the Ironbridge Gorge Museum. The Wylam engine, presumably made of cast iron, was domed at one end with the cylinder apparently included as part of a single casting.

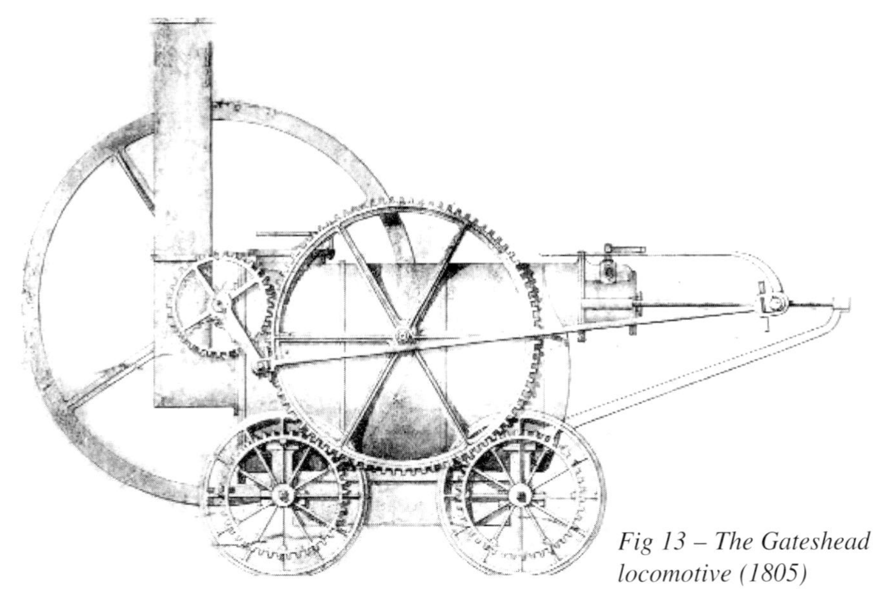

Fig 13 – The Gateshead locomotive (1805)

The Llewellyn drawing, as I have explained, cannot represent the Penydaren engine. A note added to it states that the cylinder was $4\frac{1}{2}$ inches in diameter. A single cylinder of such a size would clearly not suffice for the known performance of the Penydaren engine, though it would be quite appropriate for the much smaller 3-foot gauge engine at Coalbrookdale. Moreover Trevithick himself gives the dimensions for Penydaren as '$4\frac{1}{2}$ ft Stroake, Cylinder $8\frac{1}{2}$ inches in diameter'. The latter figure is plainly correct; the former, I think, is misleading. It has usually been quoted as evidence that the stroke of the piston was $4\frac{1}{2}$ feet; a surprisingly large figure. But Trevithick's own statements suggest a different view. For he writes: 'the engine moves forth 9 feet at every stroke'. Clearly he used the word 'stroke' in two senses; '9 feet' must indicate a two-way movement which turns the crank through a full circle; '$4\frac{1}{2}$ feet' will then refer to a single one-way movement of the piston. But like the other figure it must indicate the distance travelled by the *engine* at such a stroke; and this shows that the stroke of the cylinder, as the phrase is now understood, was not 54 inches, but 36.

For consider what happens if this is the case. The piston moves from A to B, which are separated by 36 inches. Meanwhile, the crank-pin also moves between two points at the same distance apart. But it moves in a semicircular path which measures a little over 54 inches; and it is this semicircular travel which is communicated to the driving wheels. Trevithick tells us that the engine used 1:1 gearing. Thus the driving wheels revolve in time with the crank; and they must be 3 feet in diameter to give the travel he specifies. The stroke then must be 36 inches; if it were 54, the travel would be, not 54 inches approximately, but 54 inches multiplied by 3:2, about 81 inches.

Thus I think it will be found that the '4½ ft stroake' mentioned by Trevithick not only does not refer to a 4ft 6in' stroke in the modern sense, but actually disproves it. But in fact we need not labour too earnestly to make this point. By far the simplest hypothesis is that the Coalbrookdale, the Penydaren and the Gateshead locomotive agreed in having 3-ft driving wheels and a 3-ft stroke.

Little if anything is known about the work done by the Coalbrookdale machine; the Penydaren engine, weighing 5 tons, was successful as a haulage agent but proved too heavy for the track; the standard-gauge Gateshead engine must necessarily have been heavier, at 6 to 7 tons, and was not even set to work as a locomotive but diverted at once to other uses. It had a different gear-train, giving a ratio of 2:3, and embodied a useful mechanical improvement by reversing the position of the cylinder, so that it projected from the opposite end of a cast-iron boiler. This gave better access to the fire-door; though sadly we have to disallow Tom Rolt's and John Snell's enthralling picture of a fireman at work with murderous machinery banging to and fro 'within inches of his head'! On these slow-moving engines it is most unlikely that any attempt was ever made to recharge the fire while in motion; indeed Trevithick's 1802 drawing of a road carriage clearly envisaged a stop to refuel, since no footplate was provided for a travelling fireman.

Seeing no immediate prospect of success with locomotives for heavy haulage, Trevithick turned for a short time to other projects. In 1808 he built a fourth railway locomotive, designed with a different purpose in view, namely to draw a fairly light load at a pace which rivalled that of a galloping horse. It was exhibited to the public on a circular track near the modern Euston Road, and is said to have travelled at 12mph, though Trevithick claimed, perhaps rightly, that it could have achieved 20mph on a straight and level run. Two early representations of it are known; one a broadside view printed on an advertisement card; the other a perspective view of the whole installation attributed to Rowlandson, showing the engine drawing a single passenger vehicle. This time a vertical cylinder was used with connecting rods driving directly on to crank pins on the rear axle with the wheels presumably fixed to it so as to avoid an uneven strain on the cross-head. The boiler must have contained a single straight flue, since the perspective drawing shows a crew of two, both travelling at the rear, with the chimney in front.

This novel attraction excited some public interest, and became known as *Catch-me-who-can*. But the support was not sufficient to make it a commercial success, and the exhibition was brought to an end by a broken rail, which overturned the engine. It is hard to blame Trevithick for turning his attention to other projects. In the event he never returned to work on locomotives after 1808, but plainly we should make some reference to the closing years of this remarkable man. His interests in marine engineering continued throughout the period 1805–9; in 1808, with Robert Dickinson, he took out patents for 'iron tanks for storage of cargo etc. on board vessels', and in 1809 for 'a floating graving dock; ships of iron for ocean service ... seasoning and bending timber by heated air; diagonal framing for ships; buoys made of iron' and other ideas.

In May 1810 he was struck down by typhus and gastric fever, which disabled him for six months. In September, still too weak to move, he left

Fig 14 – Trevithick's London engine Catch-me-who-can *(1808)*

London aboard a small trading vessel, which escaped a pursuit by a French man-of-war. Landed at Falmouth six days later he was able to walk sixteen miles to his house at Penponds. Meanwhile his business affairs had been neglected; Trevithick and Dickinson were both declared bankrupt in February the next year. They were discharged eleven months later; but Dickinson appears, discreditably, to have done well out of the affair.

Meanwhile Trevithick had accepted an invitation to build engines for silver mines in Peru. But his agents mismanaged them, and he himself sailed for Callao in 1816. Civil war in Peru put paid to his hopes; the insurgents destroyed his machinery, and he himself was temporarily forced to join the insurgent army. Similar troubles frustrated his engineering projects in Chile, Colombia and Costa Rica. These adventures culminated in a hair-raising transit of the Isthmus of Nicaragua. Trevithick, who could hardly swim, was nearly drowned in crossing a river, and shortly afterwards, when his canoe was upset, had a hair-breadth escape from being eaten by an alligator. His money was all spent by the time he reached Cartagena in Colombia, where he had the good fortune to fall in with Robert Stephenson. To Trevithick's disappointment he was not recognised at first by the younger man, though as he said, he had often dandled 'Bobbie' on his knee in his early years. But once identified, their meeting was cordial enough, and Stephenson gave him £50, which should have sufficed for his passage home. But apparently Trevithick was unable to find a ship sailing for England and had to go to Jamaica, by which time he was penniless again. For good measure, he was shipwrecked on his voyage back to England, and landed at Falmouth destitute except for his personal belongings and, until a kind fellow-passenger came to his rescue, with his passage money still unpaid.

His last and most extraordinary project was a design for a conical cast-iron monument 1,000 feet in height to commemorate the passing of the Reform Bill in 1832. There was to be an air lift to raise persons to the top! Trevithick's specification includes no estimate of the weight of this colossus; a mere hundred feet in diameter at its base, if not overturned by high winds it must have sunk without trace into the London clay.

Nevertheless he should on no account be dismissed as an unpractical eccentric. What could be more simple, useful and practical than his iron tanks for storing water, which proved indispensible for use on shipboard – and indeed for domestic use – superseding the cumbrous wooden cask, the unwholesome lead and the expensive copper cisterns. Countless millions of people who have never heard of Trevithick have used and valued his inventions.

4 – Murray and Blenkinsop; Blackett and Hedley

The Middleton Railway near Leeds was the scene of the next important development in locomotive history. Its authors were Matthew Murray, of the engineering firm of Fenton, Murray and Wood, and John Blenkinsop, the manager of the Middleton collieries. Conditions here were much more favourable for the use of locomotives, as the railway was heavily used, and also comparatively short, a little over three miles, so that it was practicable to re-lay it with substantial rails; and this was done in 1811–12. Blenkinsop was doubtful whether a locomotive with ordinary driving wheels would have sufficient adhesion to haul a heavy load on the awkward gradients of this line, so the rails, on one side only, were fitted with a rack to accommodate a toothed wheel. Two racks, though desirable, were ruled out on grounds of expense; and a central rack could not be used, as the railway had to leave this space free for horses.

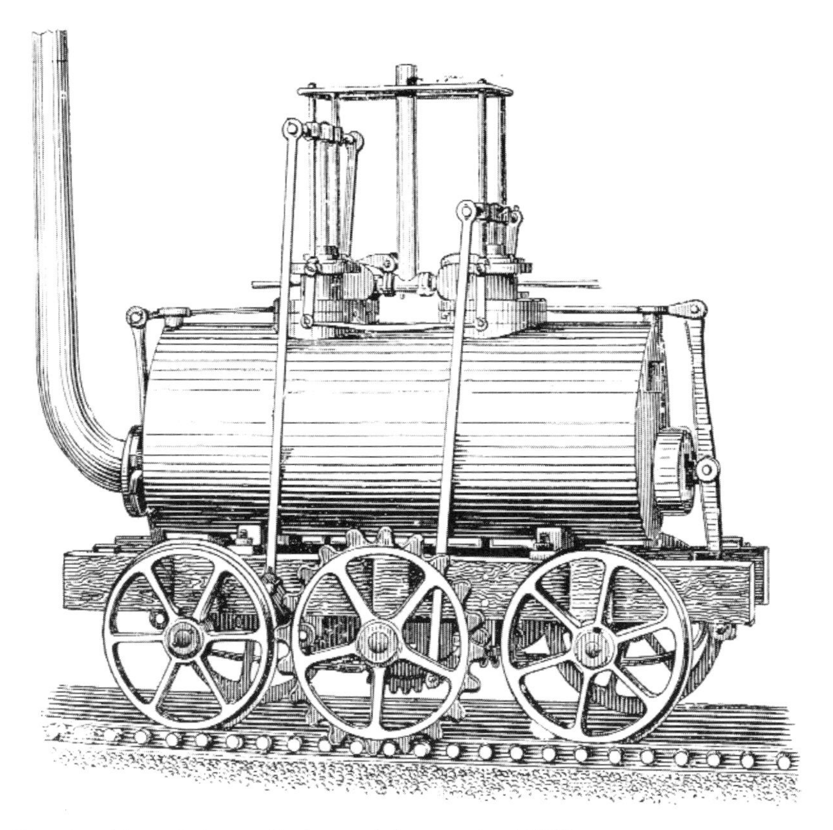

Fig 15 – Blenkinsop's two-cylinder rack locomotive (1812)

Matthew Murray built two locomotives, which were put to work in August 1812. They used two cylinders with their cranks working at right angles to one another to provide continuous power. This was not a new idea; it had been suggested by Jacob Leupold in 1725 (see page 31), and was known to Watt, though he did not adopt it. The cylinders were mounted vertically and partly within the boiler, the piston rods being guided by slide bars or rods. The drive was brought down by four connecting rods and cranks to two gear wheels, which drove a third gearwheel mounted between them on the axle which carried the toothed rack wheel; these gears of course kept the cylinders at work in their correct relationship. The two pairs of flanged wheels in front and behind served to support and guide the locomotive.

Murray did not use Trevithick's return-flue boiler, nor his system of forced draught. The latter was rather less important for a simple boiler with a straight fire-tube; nevertheless the locomotives were rather liable to run out of steam. Forced draught may have been added later, for the two locomotives, later joined by two similar ones, were still in use on the rack railway in 1835, and the quartet appear to have lasted for some thirty years in all, though possibly the rack wheel was discarded in their last years. They weighed five tons, and were able to haul a load 94 tons at 3½ mph, an impressive demonstration of their success.

The Middleton Railway still survives. Part of it became an industrial tramway, but was later closed. Part has been reopened as a tourist attraction, and carries passengers, though without any attempt to reproduce the original engines.

A second group of locomotives was built about the same time by William Hedley, who managed the Wylam colliery, west of Newcastle, for its owner Christopher Blackett. It was served by a 5-ft gauge plateway, five miles in length, which had been relaid in 1808, though not very adequately; it had an up-and-down profile and was difficult to work with horses, so that steam locomotives had to be considered. But instead of jumping to conclusions like Blenkinsop, Hedley set himself to determine by experiment whether the adhesion of smooth wheels would suffice. He built a four-wheeled trolley which could be propelled by men turning handles and tried it out in different conditions and with varying loads to increase its weight. Adhesion he found would do; a conclusion already reached by Trevithick; and this was undoubtedly the right decision. In modern practice a rack is needed only on really steep grades in mountainous country. Nevertheless his result was by no means a foregone conclusion; modern locomotives can still be troubled by slipping of the driving wheels, especially on wet and greasy rails. Yet adhesion will generally suffice, even on fairly steep inclines. The electrically-worked Bernina Railway running south from St Moritz into Italy is worked by adhesion over quite long stretches of 1 in 14; and a very short length at this inclination was to be found on the steam-worked Cromford and High Peak Railway, though admittedly the engines had to charge the bank at a run to avoid stalling.

Hedley built four locomotives for the Wylam plateway, besides one or two others whose existence is uncertain. The first, which appeared in February 1813, followed Trevithick's practice in using a single cylinder with a flywheel; but like Murray's engines it had a boiler with a single straight flue, and apparently lacked a forced draught. Hedley himself wrote in 1836: 'The (first) engine had one cylinder with a flywheel; it went badly, the obvious defect being want of Steam. Another engine was then constructed, the Boiler was of Malleable Iron, the tube containing the fire was enlarged, and in place of passing directly through the Boiler into the Chimney, it was made to return again through the Boiler to the Chimney, at the same end of the Boiler as the Fire-place was.' This was a most important improvement. The engine was placed upon four wheels and went well; a short time after it commenced working it regularly drew eight loaded Coal Waggons after it, at the rate of four to five miles per hour on the Wylam Railroad which was in a very bad state.

The return-flue boiler was a Trevithick feature which the first engine obviously lacked; but his forced draught was omitted at first, and the exhaust steam was discharged into a silencer. At a later stage, to silence it even more, it was led on to discharge into the chimney, so that a forced draught made a belated appearance.

This second engine and its two companions had a wrought-iron boiler, domed at one end and flanged at the other to take a flat cover-plate, to which the fire-place and return-flue were attached. Boiler pressure was 50 lb per square inch. So far Trevithick's influence is fairly obvious, though not acknowledged. But Hedley broke new ground by placing his two cylinders on opposite sides of the boiler near its domed end; they drove vertically upwards through a jointed connection on to two half-beams pivoted at the top of two vertical posts mounted on the wooden frame of the engine near the fire-door. From pivots placed near the centres of the two beams two connecting rods descended to drive two cranks on the same axle, which carried a central gear-wheel connected by intermediate gears to the driving axle, so that all four wheels were driven and adhesion was ensured.

Hedley's engines were robust, and worked well when the forced draught had been provided; but they proved too heavy for the lightly laid track. In 1815 he rebuilt them so that the weight was carried on two four-wheeled bogies, using a patent taken out by Chapman in 1813. In this form their appearance is familiar, since it is shown in an engineering drawing which has often been reproduced. But it gives an extremely poor idea of the actual working, as described above; while in any case some experience is needed to interpret such a drawing. For instance, it cannot distinguish two objects placed directly one behind another. Thus what appear to be two vertical posts, one large and one small, are actually four – two large and two small – placed on opposite sides of the boiler.

This account will probably satisfy most readers. But the more curious may like to explore the drawing in a little more detail. It proves to abound in the most astonishing inaccuracies. Obviously it shows two identical sets of motion, one originating at the near-side cylinder which can be seen, the other on the far side which is largely concealed. The former is shown with the piston at the bottom of its stroke, the latter in mid-stroke. I will use the letters ABC etc. to indicate the pivotal points on the near-side motion, and abc etc. for the corresponding points on the far side.

The near-side piston-rod carries a pivot A with a loose link formed of three parallel rods; it tilts slightly to the right since B moves in an arc with its centre at C. The pivot D, roughly half-way along the beam BC, carries the connecting rod DE, which extends vertically downwards, and so lines up with the crank EF, F being the axle carrying the central gear wheel, which can be seen below the wooden frame. The piston-rod is guided by the two rods GA and GH, H being the pivot on the smaller post. This was clearly an attempt to reproduce Watt's parallel motion, using the two centres G and H on opposite sides of A. But G is mounted on the connecting rod; thus its motion is governed mainly by the curve drawn by the beam tilting about its trunnion, but to a small extent by the movement of the crank-pin at the bottom. The former motion is permitted by allowing the pivot H to slide to and fro (the extended top surface of the smaller post confirms this idea); otherwise the motion would seize up at once. As to the latter, the upward and downward movements of the rod will follow different paths; this feature remains as a troublesome irregularity.

So far, I think, the drawing is to be trusted, as correctly showing an extremely clumsy piece of engineering on Hedley's part. The case is very different when we come to the far-side motion, where the drawing contains some astonishing mistakes; for (1) the far-side connecting rod is shown attached to the *near-side* beam! – though this only appears clearly in a good reproduction. A trace of it is also shown on the far-side beam at *d*, but this points vertically downwards instead of slanting to the right, as it should; (2) the slanting rod goes nowhere; it should end at a crank-pin *e* on a level with the axle F*f*, and just visible below the wooden framing to the right of the gear wheel; and (3) a pivot should be shown at *g*; this carries the rods *ga* and *gh*, which line up with one another at mid-stroke; thus *a* should be shown directly behind B. The drawing, by omitting this pivot, converts the motion into a rigid framework of rods which would prevent the engine from moving at all!

Let us now leave this intricate detective work and return to the engines themselves. As we have seen, they were mechanical curiosities, but they seem to have worked well enough, largely because of their excellent boilers. Hedley at least thought them more efficient than Stephenson's, which may well be true at this period. They continued at work in this form for ten years or so, but some time between 1825 and 1830 the plateway was altered to an edge railway, still retaining its unusual 5-ft gauge. The *Billies* then reverted to their four-wheeled condition, as allowed by the stronger rails; and the opportunity was taken to remove their eccentric linkage and substitute an orthodox version of Watt's parallel motion. The wooden posts disappeared, and were replaced by iron frameworks. The two at the rear end adjacent to the cylinders were rigid and were braced by diagonal stays. The other two,

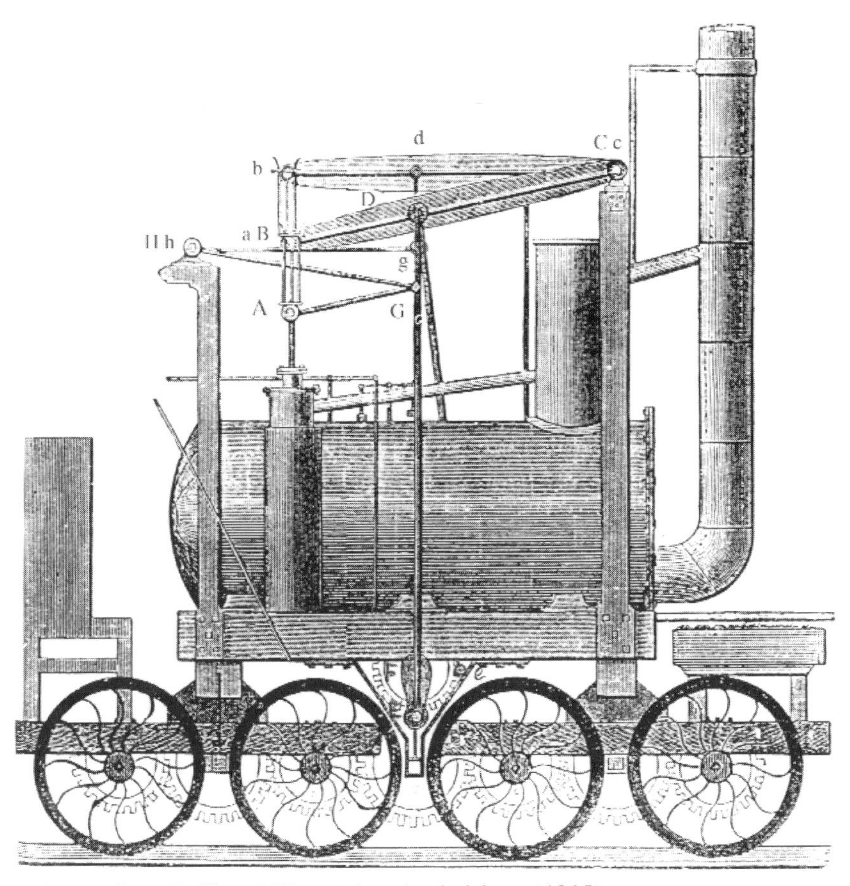

Fig 16 – Hedley's Puffing Billy*: eight-wheeled form (1815)*

Fig 17 – Hedley's engine as rebuilt (1825–30)

which supported the beams, were arranged to tilt fore-and-aft from their bases. The piston-rods could thus be connected directly to the beams; and the neccesary guidance was provided by rods, pivoted at both ends, connecting the rigid rear-end structures with points about half-way along the beams, just forward of the connecting-rod bearings. This fulfilled Watt's basic principle of two arcs of equal but opposite curvature which cancelled out and guided the piston-rods in the required vertical path.

In this form the *Billies* soldiered on efficiently for more than 30 more years, till in 1862 the Wylam railway was converted to standard guage. *Puffing Billy* was then taken out of service and went to the Science Museum in London; her sister engine *Wylam Dilly* apparently continued until 1867 and is now in the Edinburgh Museum.

5 – Some important improvements

We must now break off this narrative and explain some fundamental points of locomotive design as they emerged: the layout of the cylinder; the valves for admitting and exhausting steam; expansive working; and the mechanics and functions of lap and lead.

The type of cylinder most familiar to the general public is that used in motor-car engines. It is open at the bottom, and has two valves at its upper end, one to admit fresh vapour, the other to discharge the waste products to the exhaust. Railway enthusiasts will not need to be told that the steam locomotive uses double-acting cylinders, as introduced by James Watt in 1775. These have a single port at either end; but each port does double duty; it admits live steam at the beginning of each stroke, but at the end, or in practice a little earlier, it opens to allow the expanded steam to escape from the cylinder as the piston returns. Thus each port has to be connected alternately to the live steam supply and to the exhaust pipe through which the used steam is discharged. Various forms of valve gear have been devised to make these connections at the appropriate moments; for it was soon found that precise timing was of the utmost importance for the efficient working of the locomotive.

The great majority of steam locomotives in their final form, which was reached in the late twentieth century, made use of the slide valve or its more recent equivalent, the piston valve. We have just been reviewing the period in which the slide valve first came into use; it was patented by William Murdock in 1799 and introduced by Matthew Murray in 1802. But the earliest slide valves were crude affairs; vital improvements such as lap – to be explained later – were badly designed or even ignored. Thus during the first twenty or thirty years of the nineteenth century the slide valve was beginning to be adopted and improved; but older types of valve were still in use and will need to be described.

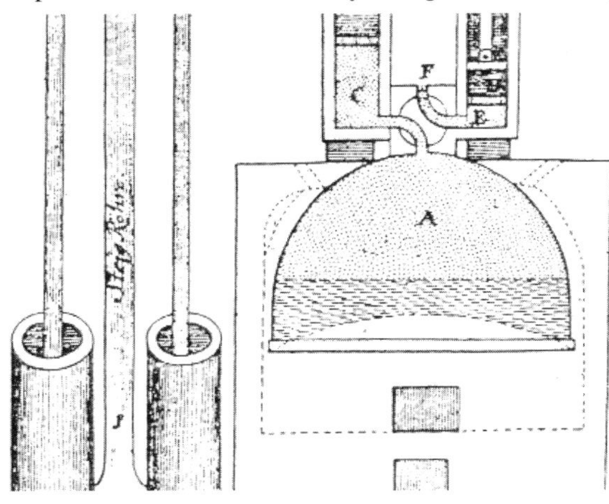

Fig 18 – Leupold's steam valve (1725)

A very simple device used in some eighteenth-century and early nineteenth-century engines was the semi-rotary plug valve. This is shown in a drawing by Jacob Leupold in 1725, though here it served to govern the steam admitted to two single-acting cylinders rather than to the two ends of a double-acting one. It consisted of a cylindrical plug fitting closely into a socket, like the tap of a cider-barrel. The socket had four openings, two of them leading to ports at opposite ends of the cylinder, the other two connected to the live steam supply and to the exhaust. The plug contained two L-shaped passages, and could be turned through a right-angle to distribute the steam as required. Its obvious limitation was that steam had to be admitted at one end of the cylinder or the other throughout the whole of each stroke; there was no possibility of expansive working: of economising steam by cutting off the supply part-way through the stroke, and allowing it to expand in the cylinder. Crude as it was, this device had a long life. It was used by Trevithick in 1803, though not in a locomotive, and is believed to have been used by Hedley in 1813 on the Wylam engines in their original form.

An obvious improvement on the primitive rotary valve was an apparatus that could permit of expansive working. Watt was well acquainted with this possibility, and drew a diagram showing the gradual diminution of the pressure in a cylinder when the steam supply is cut off early in the stroke; but with the very low initial pressure used in his engines he reckoned that the benefits of expansive working were not worth having. Trevithick's practice was flexible; in some of his engines he used the rotary valve; and even when his valves allowed of expansive working, he sometimes preferred full pressure throughout for heavy work. Other reports mention cut-offs at 2:3, 1:2, 1:3, 1:4, and even 1:9 of the stroke! But some of these of course could be used only on stationary engines, not on locomotives.

Expansive working called for an independent admission valve for the steam. This could be operated very simply by a rod running parallel to the piston-rod and equipped with tappets which were struck by the cross-head or some similar projection; the to and fro motion of the rod opened and closed the steam valve. The rod in question is clearly visible in the Llewellyn drawing of the Coalbrookdale locomotive, including a handle at the end to allow the valve to be opened and closed when the machine was at rest.

But this type of valve shared one of the limitations of the plug valve; its movements were produced by the impact of a tappet on a lever. This had the advantage that the valves could be opened and closed very quickly; the drawback was that it could not be made large enough to allow a free flow of steam at high speeds. A large valve would necessarily be heavy; and no engineer could rejoice at the thought of levers and tappets being subjected to violent shocks, if a large and heavy valve had to be suddenly shifted. The great advantage of the slide valve was that it moved smoothly to and fro, in the-manner of a pendulum, or indeed of the piston in the cylinder. It could thus be easily enlarged to allow of ample steam passages. But in its crude, original form it opened very sluggishly, so that high-speed working was ruled out. The history of the slide valve shows how this defect was overcome, and the smooth harmonic movement adapted to provide a rapid opening of the steam ports. In a sense the design is well known, since almost every book on locomotive engineering contains a sectional drawing in which the valve and steam passages are viewed from the side. But this drawing conceals one vital feature, namely the generous size of the steam passages. It shows the route by which the steam enters the cylinder and passes to the exhaust pipe when its work is done. But the uninstructed reader might be forgiven for thinking that the steam passages are circular or square in section. In point of fact their cross-section is a long narrow rectangle seen from its narrow end. What the drawing conceals is their extent measured along the reader's line of sight; this hidden dimension may be some nine or ten times the breadth revealed in the drawing. Thus the cross-section of these passages is far greater than it appears to be; and they enter and leave the cylinder by curving slots extending nearly one-third of the way round its circumference.

All this could easily be shown by a model; less clearly by a perspective drawing. The best we can do here is to describe the whole apparatus as seen in the round.

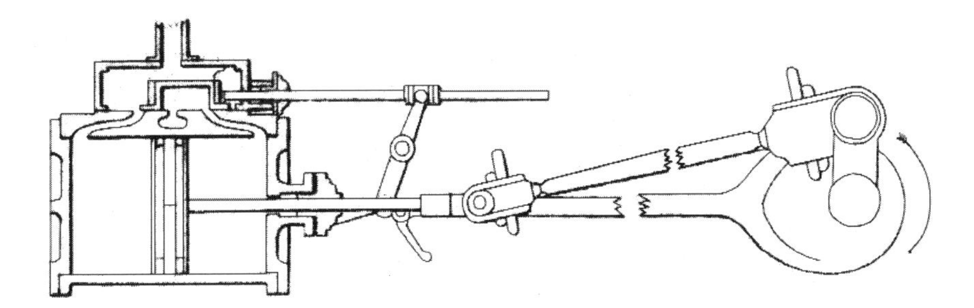

Fig 19 – Sectional drawing of slide valve

The slide valve itself has a form resembling a shallow oblong pie-dish turned upside down. It is driven by a rod, and slides to and fro on the valve-bed, a perfectly flat surface machined on top of the cylinders. A cover is bolted to the valve-bed large enough to allow the valve to slide beneath it, and equipped with a steam-tight opening through which the valve rod can pass. The space between the cover and the bed is called the steam-chest, and it can be enlarged as required to provide a good reservoir of steam. The pressure of the steam keeps the valve firmly pressed down on the bed, though of course free to slide along it.

For simplicity's sake we have described the usual layout, in which the valves are placed above the cylinders; but other dispositions are possible, either with the valves below the cylinders, or placed between them and sharing a common steam-chest. 'Up' and 'down' in our present description therefore mean 'away from the cylinder' and 'towards it.'

The valve bed accommodates three steam ports, the outer two communicating with the two ends of the cylinder, the central one with the exhaust pipe. In shape these ports are long narrow rectangles, the large dimension extending across the valve bed from side to side, and so invisible in the drawing. At one end of its stroke the whole valve slides clear of one of the outer ports, so that live steam can pass from the steam chest into the cylinder. Atthe same time the hollow space formed in the valve is shifted across, so that the exhaust steam can pass into it from the other port and reach a passage opening out sideways in the cylinder block. The central space itself is always covered by the valve, to prevent live steam going to waste; when one port is open to live steam the other communicates only with the exhaust pipe.

We have described the slide valve as going to and fro with a smooth harmonic motion resembling that of a pendulum, which gathers speed only by degrees as it moves away from its extreme position, to reach a maximum at mid-stroke. This motion of the slide valve, or of its later development the piston valve, is derived from other moving parts by various mechanisms, or valve gears, to be described in due course. Our present concern is to explain how the smooth motion of the slide valve can be used to provide the rapid opening and closing of the ports which is required for efficient working.

In the earliest designs the 'feet' of the valve, its sliding surfaces, were made only just long enough to cover the ports; hence the valve began to open at the same moment as it began to move; and this initial movement is very sluggish, as we explained. But the whole picture was transformed by the introduction of 'lap'. The lap is the distance by which the feet of the valve, in its central position, extend extend beyond the limits of the steam ports. Though this distance is quite small, it is traversed by the valve when it is moving very slowly away from its extreme position. By the time it begins to uncover the ports it will have built up speed; thus the full opening of the port will be quickly completed.

The valve therefore begins to move before the piston begins its stroke, after which both move in the same direction. The eccentric, which moves the valve, has to be adjusted to allow for this advance movement, and in modern practice the valve moves about half a stroke in front of the piston's movement.

Expansive working was possible with this type of valve; but initially the point of cut-off had to be fixed at the time of construction. In practice it was a compromise; a cut-off at about 60% of the stroke enabled the economy of expansive working to be realised in some degree while avoiding the danger that the engine might stop in such a position that neither cylinder could provide enough power to restart it. This problem indeed still arises; but modern locomotives can easily be reversed, and the once familiar process of 'setting back' for a stroke or two enables them to stop in such a position that they can start forward again.

So far we have been assuming that the valve, moving rapidly, begins to uncover the port at the exact moment when the piston is ready to begin a new stroke. But it is possible to introduce a further refinement called 'lead', which allows the steam to be admitted a little before the piston reverses its movement. The steam admitted in front of the piston as it nears the end of its stroke helps it to come to rest and restart. This momentary back pressure in front of the piston is of course quite different from the back pressure produced by an inefficient exhaust system, which begins quite early in the stroke, resists the action of the live steam on the other side of the cylinder and greatly reduces the efficiency of the engine as a whole.

Since lap and lead are often mentioned together, it is worth mentioning an important point in which they differ. The lap is a basic constructional feature, and has to be fixed when the engine is designed; it could only be altered by fitting new valves. But the lead could easily be altered if the eccentrics were reset, though in full-size practice they are seldom designed to allow this. In any case, the lead is normally constant with Walschaerts gear, but varies helpfully with Stephenson's. Swindon practice was to arrange for negative lead in full gear (helpful for starting and heavy work at slow speeds) and increasing positive lead as the cut-off was reduced and piston speeds increased.

This discussion of lap and lead applies mostly to a later period than the early nineteenth century. Although lapped valves were invented quite early in the century, they were actually adopted rather slowly; valves without lap were still in use as late as the 1830s, and can be seen in old sectional drawings; and where lap was used, it was often too short. Valves set with lead are a later improvement.

A former chapter dealt with the adhesion of the driving wheels, which may not be sufficient on wet or greasy rails. In this connection it is worth noting that the use of two cylinders, though an enormous improvement on only one, cannot ensure a perfectly constant tractive force. One reason for this is fairly obvious; with expansive working, the pressure of steam in a cylinder falls away in the latter part of the stroke; and it is clearly impossible that the other cylinder should gain pressure at a rate that would exactly compensate for the loss.

A second reason is that an assemblage of connecting-rods and cranks is inherently incapable of transmitting a steady impulse to a rotating axle. Thus in a locomotive at the 'dead-centre' position, with the piston-rod in line with the crank, the thrust on the piston-rod will be negligible, if any at all. But if the crank moves on, say by 10°, the thrust will already be considerable if the engine is well designed, yet the leverage will be so unfavourable that only a small part of it will be converted into useful work; the rest will be absorbed by forces tending to distort the motion and, more important, by shocks imposed on the axle bearings. In this respect the outside-cylinder locomotive has a slight advantage, in that the thrust is delivered close to the point where it is needed, at the rim of the wheel; with inside-cylinder engines the connection with the wheel rims is relatively indirect, but that with the axle-bearings direct; so the shocks are directly transmitted to the bearings.

A third point is less important, but is intriguing enough to deserve notice. Imagine the position when the piston is half-way along the cylinder. At this moment the connecting-rod will be at right angles to the crank. But clearly, since the connecting-rod is out of line with the piston-rod, the piston has to move a little further before the crank stands at right angles to the piston-rod. In effect, when the piston has traversed half its stroke, the crank has moved round about 85°, leaving 95° to be covered in the second half of the stroke.

This means in effect that a half-turn of the crank-pin when it is nearer the cylinder gets more than a half-share of the piston's effort, and the other half-turn less. If we use the letters A,B,C,D to stand for successive half-strokes of the piston, the pressure of steam in the cylinder will be high in A and C; low in B and D. But considering the leverage of the connecting-rod on the crank-pin, it is A and D which enjoy the advantage of the crank-pin being near the cylinder, while B and C lose out. To summarise:

A: pressure high, leverage good.
B: pressure low, leverage poor.
C: pressure high, leverage poor.
D: pressure low, leverage good.

Using four cylinders does nothing to iron out these variations, since it provides two pairs of cylinders moving in unison. But three cylinders will pretty certainly give a more even turning movement, though the mathematics are far too complex to be treated here. At all events three cylinders are extremely desirable where adhesion is a problem, and have been used with great advantage on the 'Schools' 4-4-0s of the one-time Southern Railway. The much-lamented 'turbomotive' 4–6–2 of the LMSR of course delivered a perfectly even thrust, and probably deserved a better reception.

6 – George Stephenson at work

Fig 20 – George Stephenson (1781 – 1848)

George Stephenson was poorer and less distinguished in his origins than any of the engineers we have so far considered. His father, Robert Stephenson, worked as a fireman, tending the furnaces of pumping engines at the mines, and moving from place to place wherever work could be found. George was born at Wylam, six miles west of Newcastle, on 9 July 1781, the second in a family of six children, with three competent brothers and two aimiable sisters. At first he worked on a farm, but he soon migrated to unskilled work at the coal-pits. His gifts began to be recognised; he was enterprising, imaginative and observant, and rapidly acquired a remarkable grasp of the workings of machinery. In 1798, at the age of seventeen, he was put in charge of a new engine-house at West Row, where his father was still working as a fireman. In 1801 he obtained a more responsible position as a brakesman at Willington Quay, in charge of the pit-head winding engines. His superintendent here was Robert Hawthorne, who had recognised his ability and gave him valuable help and advice. He married the next year, and his son Robert, his only child, was born in 1803.

For the next eight years his progress was steady, though not dramatic; by economy and hard work he was able to provide his son with a good education, an advantage which he himself almost wholly lacked. His great breakthrough occurred in 1811. He had by this time acquired quite a reputation for ability; he was invited to go to Killingworth, five miles north-east of Newcastle, and rebuild an old engine of a design much like Newcomen's, which had totally failed to lift the water from a flooded pit. Stephenson identified its various deficiencies and set to work. His rebuilding was a resounding success; indeed for the first few strokes the engine's vibrations shook the engine-house to its foundations; but it soon settled down, as he predicted it would, and by the next day the pit was dry.

Next year at Killingworth a post as engine-wright fell vacant, owing to the accidental death of its occupant, and Stephenson was appointed in his place. The pit was one of a number owned by the so-called 'Grand Allies', the largest employers on Tyneside; Stephenson was at once entrusted with the care of all the machinery at their collieries at a salary of £100 a year, a very handsome figure for those days, and with considerable liberty to take on other work. Moreover his Killingworth appointment had the great advantage that it brought him into close contact with Nicholas Wood, who had become 'viewer', or manager, at Killingworth at the early age of seventeen. Fourteen years younger than Stephenson, Wood soon developed into a mining engineer of great distinction, an expert on every aspect of colliery working, including its transport requirements. Their friendship undoubtedly helped Stephenson to expand his activities and profit by new experience.

Meanwhile, after a slack period, interest in the steam locomotive was-

reviving, as the price of horse fodder had rocketed up owing to the war with Napoleon, and an alternative means of haulage had to be found. Trevithick's work was well known on Tyneside, where he had built his so-called 'Gateshead engine' in 1805 for Christopher Blackett. Away to the south, near Leeds, Blenkinsop and Murray had shown that locomotives could work successfully, while on Tyneside William Hedley was building locomotives of a quite different design for the wagon-way at Wylam. Meanwhile Stephenson's responsibilities both towards the Grand Allies and others had rapidly expanded. He was the obvious choice to build them a locomotive; indeed it seems that Stephenson himself made the proposal to his employers. The result was the *Blucher*, which started work on 25 July 1814.

Blucher was a close copy of the Middleton engines with no attempt to follow Hedley's practice, except that Stephenson dispensed with the rack wheel and its expensive special track. The engine ran on flanged wheels over cast-iron edge rails; it had a straight-flue boiler 8ft long and only 3ft in diameter, like the wheels. There was a plain flue-tube 20 inches in diameter, with no attempt to provide forced draught. This was obviously cheaper and simpler to construct than a return flue, though much less efficient; but allowable where coal was cheap. The cylinders were connected to one another and to the driving axle by no less than five gear wheels, with the drive coming alternately from the two ends. But *Blucher*, though noisy and troublesome to maintain, seems to have worked well enough; it hauled a thirty-ton load up a very slight rising grade (1 in 450) at four miles an hour; though possibly this was not much better than Trevithick's Penydaren engine.

The gear wheels, however, soon showed signs of wear, and a second engine was built in 1815, incorporating some important improvements. It had a much larger boiler, apparently 4ft in diameter. Forced draught was provided, the gear train disappeared, and the cylinders, again set vertically in the boiler, were spaced out to lie directly over the driving axles. The drive was taken through the usual cross-heads and connecting-rods to crank-pins on the wheels. Trevithick had used this method on his London engine; but the use of two cylinders introduced new problems, as both they and their crank-pins had to be out of phase, making it impossible to use connecting-rods of the usual modern type. After an unsuccesssful experiment with cranks, Stephenson came up with a new idea; this was to connect the two driving axles by an endless chain running over sprocket wheels. In its form the chain, once again a new idea of Stephenson's, resembled that used on modern bicycles. The engine was unsprung.

With its large boiler *Wellington*, the new engine, must have been much heavier than *Blucher*, its predecessor, and Stephenson soon had to meet complaints of broken rails. A partial solution was offered by a patent taken

Fig 21 – Stephenson's second locomotive, February 1815

out, in association with William Losh in November 1816, for a new design of cast-iron rails, having diagonal joints at the ends to provide continuous support for the wheels and reduce the shocks set up at the rail-joints. Nevertheless from 1816 onwards the locomotives were given a form of springing covered by the same patent. Springing was certainly not an unmixed blessing with the cylinders placed vertically, as they were, for the up-and-down motion of an engine on springs was apt to interfere with the free movement of the pistons. On the other hand, if a shock occurred when the crank-pin was at the top or bottom of its path, it could not relieve the strain on the connecting-rod by either quickening or slowing its pace; it was precisely then that springing was most required. But in 1816 it was not possible to obtain plate or leaf springs, as used today, which would support the weight of a locomotive. Stephenson met the problem by using 'steam springs'; the engine's weight was carried by vertical cylinders filled with steam. Though messy and troublesome this arrangement was successful, and was adopted on the next few locomotives – including an 0–6–0 built in 1816 for the Kilmarnock and Troon railway in Scotland – until adequate leaf springs became obtainable in 1825–8. Several engines of this general type were in use on the eight-mile Hetton colliery railway, or at least on its level sections. One of these survived until 1908, and this may well be the engine illustrated in the *Pictorial Encyclopedia of Railways,* p22; it can hardly be *Wellington*, if E W Twining's model of it is correct, since that engine seems to have had a much longer boiler and wheels set further apart.

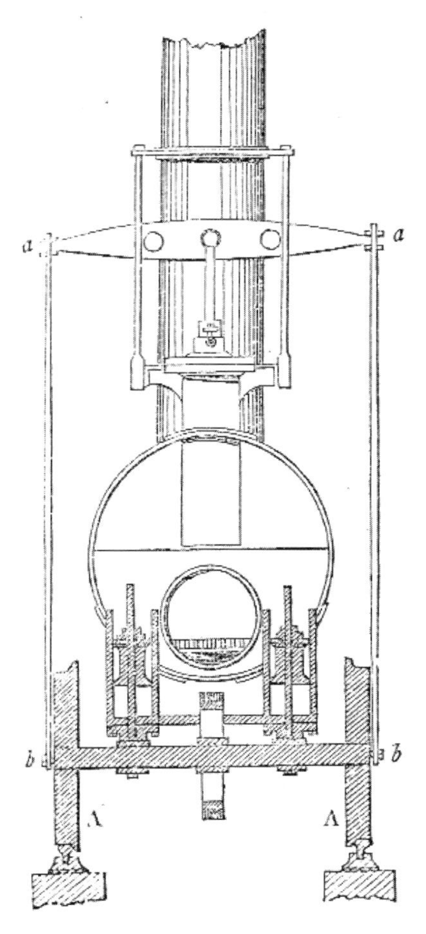

Fig 22 – Stephenson's third locomotive, with steam springs

The chain coupling was soon given up and replaced by connecting rods, though this involved using return cranks on one of the axles, so that the crank-pins of the coupling rods could revolve in time with each other although those of the connecting rods were out of phase. It was difficult to make this device strong enough to resist the forces it had to transmit; but it was provided on the *Locomotion* of 1825, and worked well enough to be fitted to the earlier engines, replacing the chain coupling with which they were built.

All these locomotives were equipped with slide valves worked by a single slip eccentric on each axle. The drive for the valves was led through cranks to rods ascending almost vertically up one side of the boiler, and thence by cranks to a transverse axle to reach and actuate the valves.

With the aim of improving locomotive performance on the Killingworth line, Stephenson and Wood carried out some tests in 1818 to determine the sources of resistance. Stephenson identified three types, namely the friction between axles and their bearings, the friction of wheels on the rails, and the effects of gravity caused by inclines. As noted by Samuel Smiles, the scientific theory had been developed long before by Vince and Coulomb, but was unknown to practical engineers. Stephenson had to devise his own methods, and built a dynamometer which measured the pull by its extension of a spring, the method still in use today. He reached the rather unexpected conclusion that frictional resistance does not vary with the speed of motion. But he also reasoned that the lines to be used by locomotives should be made as flat as possible, with inclined planes worked by cable haulage interposed where rises and falls were inevitable. In so doing he was of course closely following the practice of canal engineers, who had to use level stretches interrupted by locks. This layout was followed by Stephenson and by Brunel, whose main lines dispensed with the complications of cable haulage, but still used long stretches either on the level or very slightly inclined, interrupted by short steep gradients where assistance could be provided. Stephenson's Liverpool & Manchester railway had two such gradients on either side of the Rainhill level, and Brunel's London to Bristol main line was similar, with the uphill slope between Bath and Swindon concentrated in two lengths at 1 in 100 at Box and Dauntsey. Meanwhile Joseph Locke rightly preferred long stretches at a moderate but even inclination; a particularly good example is the London & Southampton main line, with long slopes of 1 in 250 or less; the London & Birmingham used 1 in 330. Others argued, fallaciously, that steep gradients are no obstacle, since the time lost on the ascents can be made up by high speed running downhill. Mathematics proves them wrong. At 60 mph 10 miles takes 10 minutes; 5 miles at 30 mph will also take 10 minutes, and the remaining 5 miles are bound to take some time, however colossal the speed.

Stephenson had been very active during these years in designing, building and maintaining a total of about sixteen locomotives; but this was by no means the full tale of his activities. He made extensive improvements to the handling and transport of coal both below ground and on the surface. Tram rails were introduced, with switches and crossings, and horses were replaced by stationary winding engines where locomotives could not be used. These included the underground engines at Killingworth, known as *Geordie*, *Jimmie* and *Bobbie*, after George and his brothers. And other collieries soon benefitted from his methods and advice.

The Hetton colliery railway was laid out by Stephenson between 1818 and 1822, following the principles just mentioned, with level stretches interspersed with cable-worked inclines. It ran northwards from Hetton-le-Hole,near Durham, to reach the River Wear near Sunderland, a distance of eight miles through very hilly country, with a summit level of nearly 650ft at Warden Law. Stephenson did it all; surveyed the route, constructed the necessary bridges and earthworks and built the locomotives bringing his total so far to about sixteen, a fine proof of energy and versatility combined.

Particularly important was his design of a safety lamp for use in mines where explosive gas, or fire-damp, was present. The need had been horrifically shown by a disastrous explosion at Felling pit, near Gateshead, in 1812, when 90 men and boys died. Sir Humphry Davy, well known for his work on chemistry and electricity, was invited to devise a solution and visited the mines in August 1815. But Stephenson was also concerned, and had already taken part in a hazardous and successful operation to extinguish an underground fire as early as 1809. He presumably knew of Sir Humphry's visit, but Davy clearly had no idea that the remote North-country mechanic was already engaged on the problem. Davy went quickly to work and announced a successful result on 9 November. Meanwhile Stephenson had tested his first lamp on 15 October, and a second on 4 November; the third and final version was ordered on 20 November. His lamp and Davy's were very similar in construction; both were fitted with a metal surround to protect the flame, Davy using wire gauze and Stephenson at first a perforated plate; Stephenson also retained a glass chimney, which Davy dispensed with. Davy's lamp was soon adopted all over the world; but many North-country pitmen still held that the *Geordie* was safer, and in some conditions it seems that this was true, as was proved by an explosion in 1825 where a Davy lamp was in use.

Meanwhile an unfortunate controversy had broken out by 1817 as to who should claim priority for the invention. Davy's work had been immediately acclaimed, but he unwisely attempted to deny Stephenson any credit, on which his supporters, including the Earl of Strathmore, made an effective rejoinder. The general public naturally awarded sole credit to the much better-known Sir Humphry, and he received a handsome present of £2,000 for his invention of the safety lamp, while a consolation prize of 100 guineas was accorded to Stephenson. However his North-country friends held a dinner in his honour and presented him with a handsome silver tankard and a further prize of £1,000.

This should have been the end of the controversy; but Davy persisted, ungenerously and unfairly, in claiming sole credit, although Stephenson had certainly been at work longer and incurred far more personal danger by extended trials in the pits themselves, bringing a naked light into galleries where explosive gases were known to be present. The just account took some time to emerge.

Some further light is thrown on Stephenson's activities in 1819, when the unrest arising from the Peterloo massacre aroused widespread fears of a general insurrection. 'I'm sorry to inform you,' Stephenson wrote, 'that I have become a soldier. We send a dozen every day to Mr Brandling's to learn exercise, and I do assure you we can handle the sword pretty well ... But I think if any of us be wounded it will be in the Back. Three hours' drill daily, and plenty to eat and drink at Gosforth Ho. The Reformers have also been learning their exercises. I do assure you they have alarmed the Gents in our country, especially our worthy masters. Ld. Strathmore's cavalry having marched through Winlaton lately, the day following the Reformers of that place marched their cavalry through in imitation of the Noble Lord's. It consisted of 72 asses with hardy nailers for their riders.' Stephenson evidently took the side of his employers, and of the established order.

7 – Locomotion

Locomotion is perhaps the best known of the early railway engines, more striking even than the *Rocket* in its archaic appearance and construction. And it survives virtually unchanged, whereas the *Rocket* was largely rebuilt. Before describing it, something must be said about the Stockton & Darlington Railway for which it was constructed, the first steam-using public railway in England.

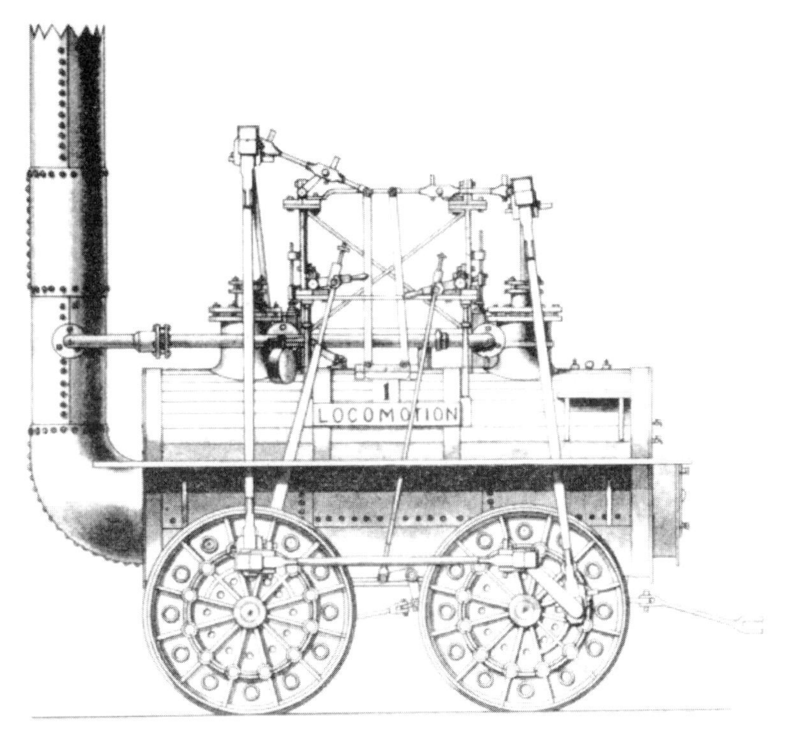

Fig 23 – Locomotion, *1825*

Like most early railways it was built to transport coal from the mines, though it soon came to carry other traffic. The Tyneside collieries had enjoyed the advantage of having a navigable river close at hand; those around Bishop Auckland were less fortunate. The River Wear at this point is fast-flowing and shallow; a well-known photograph taken at Durham, fifteen miles downstream, shows the Cathedral and the Castle with the river, still shallow, flowing over a weir. The coals would have to be carried eastwards to reach deep water on the Tees, quite close to its mouth near Stockton, before they could be shipped away.

The principal colliery at Witton Park lay about 450 feet up, near the Wear valley to the west of Bishop Auckland. The shortest route would have to rise to a summit level of some 650 feet at High Etherley, and then descend nearly 300 feet to cross the valley of the River Gaunless, a minor tributary of the Wear, near St Helen Auckland. It must then rise again to 450 feet to Brusselton summit before descending again to Shildon, near the later settlement of New Shildon, where locomotive working could begin. Thus four cable-worked inclines were needed west of Shildon, where the workshops were established, and later engines designed and built.

The length intended for locomotive working now ran to a point near the modern settlement of Newton Aycliffe. From here it would have been easy to continue eastwards to Stockton; but there were obvious advantages in taking the line close to Darlington, so it swung round southwards parallel to the Great North Road (now the A1), and the later North Eastern Railway, to pass just north of the town, which was served by a short branch. The main line continued eastwards over what became the well-known level crossing with the main line to Edinburgh before turning north-east to reach Stockton. The distance was thus some 25 miles, as against 16 by the shortest possible route.

But this route was only settled after lengthy deliberations. A canal had been proposed as far back as 1768, but the scheme was unworkable, and by 1810 there was a counter-proposal for a tramway, backed by the influential Edward Pease. At a subsequent meeting John Rennie was commissioned to report, and once again recommended a canal. The decision is difficult to understand; whatever the route, it would have been impossible to avoid at least one long flight of locks, which could not fail to cause lengthy delays.

In any case, the promoters' opinion was swinging back to some form of plateway or tramway; no doubt the successful operation of Stephenson's inclined planes and locomotives on Tyneside and at nearby Hetton was already known. Thus in 1818, when a canal was once again proposed, Edward Pease and his allies protested. The promoters were won over, and agreed to commission a survey for a tramway or railway. They applied for parliamentary powers in 1819; the bill was thrown out by a small minority. A second attempt succeeded on 19 April 1821; the very same day George Stephenson and Nicholas Wood went to call on Edward Pease.

Pease was impressed by Stephenson's air of competence and experience, and his straightforward unaffected manner. There was such an honest sensible look about George Stephenson, and he seemed so modest and unpretending, and he spoke in the strong Northumberland dialect. Pease himself was a man of some education, but his diction also reflected his Quaker allegiance, invariably preferring 'thee' to the more fashionable 'you'. Between them they succeeded in persuading the Board to decide for an edge railway, in preference to a plateway, and Stephenson was commissioned to make a survey, revising and improving the route previously recommended by George Overton. Favoured by excellent weather, the survey was rapidly completed in the autumn of 1821, and the plans, estimate and report were presented the following January. The charge for the survey was the modest figure of £140, and the estimated cost of the new line was a little under £61,000, about £2,400 per mile.

Stephenson had time now to consider the provision of locomotives, in consultation with his son Robert, who had already assisted him on the survey, and who founded the well-known locomotive building works in 1823. The locomotives in use at Hetton and elsewhere followed the pattern adopted in 1816, with vertical cylinders working twin connecting rods, chain coupling of the driving wheels, and steam springs. There had been gradual improvement in performance, but speeds seldom exceeded 10mph, and something more versatile was called for on a line expected to carry passengers as well as goods. Drawings have now been discovered which show that several new plans were considered before *Locomotion* attained its final form. It followed George Stephenson's established practice in its main conception but had some interesting new features.

It retained Stephenson's original design of boiler with a single straight flue. This was by no means efficient, but it was simple and cheap to construct. The violent rush of gases through the flue to the blast pipe was sufficient to sustain a brisk fire, at the cost of throwing much red-hot coal out of the chimney. The waste of fuel did not matter too much in a district where it was cheap and plentiful; the rain of hot cinders on the passengers' heads had simply to be endured.

There were three main innovations. First the steam springs were abandoned. It was a year or two before reliable plate springs became available, so the engine was unsprung. Perhaps it was thought that a newly laid track could be made sufficiently firm to accept unsprung locomotives without inconvenience; expense and maintenance trouble could thus be avoided, and there would be no interference with the regular motion of the pistons. This again should have led to a great improvement in the efficiency of the cylinders; if no extra space had to be allowed at the two ends to ensure that the pistons did not strike them, the length of the cylinders could be tailored to fit the working stroke of the engine, and a great waste of steam avoided. This single fact may have been enough to account for *Locomotion's* advantage in power over Stephenson's earlier machines; though it may have ridden very roughly as the track began to show the effects of wear and distortion. Next, the chain coupling was discontinued. It was of course impossible to use connecting rods without special provision, since the two cylinders and their crankpins were not moving in phase. But by now Stephenson was sufficiently confident of his manufacturing skill – or his son's! – to use a return crank, which set up a secondary coupling pin at an angle of 90° from the main crank-pin, and enabled connecting rods to be used. Hackworth had apparently designed this innovation.

The most remarkable departure from precedent, however, was the revival of parallel motion in preference to the slide bars which Stephenson had invariably used hitherto. This has been condemned as a retrograde move; but a possible reason for the change may be suggested. The railway was intended to carry passengers, and the Stephensons must have envisaged higher speeds than the 10 mph or so already achieved; in fact *Locomotion* on her trial run attained 25 mph with a very considerable load. Lubrication was still something of a problem; and it may have been thought that cross-heads moving at high speeds along slide-bars would overheat and jam. A guidance system provided by pivoted rods would be far more complex, but the lubrication of the pivots would present no problems, as the friction would be small.

At all events, parallel motion was quickly, though briefly, adopted, by other designers. Besides *Locomotion* and her sisters, four engines in all, it appeared on Hackworth's *Royal George* of 1827; on Hedley's rebuilds of the *Puffing Billies*, some time between 1825 and 1830; on at least two half-beam engines, the *Stourbridge Lion*, by Foster and Rastrick, which went to America; on the very similar *Agenoria*; and on Marc Séguin's eccentric 0–4–0 in France.

To the casual onlooker *Locomotion* presents a bewildering tangle of rods on the top of its boiler. Four sets of parallel motion had to be provided, to guide both ends of the two cross-heads. The piston-rods, cross-heads and descending connecting-rods are easily identified. Between them lay a fixed rectangular structure of rods with diagonal bracing. Its purpose was to provide fixed points for the pivots of four rods which pointed inwards towards the centre of the engine. These again were pivoted to four much longer rods which swung from pivots quite close to the centre. Obviously these pivots could not be fixed, as their distance from the intersection with the shorter rods must vary; they were carried on four nearly vertical rods pivoted on top of the boiler. This allowed the upper pivots to swing to and fro for a few inches as the rods assumed their varying positions, The outer ends of the longer rods were attached to the cross-heads, and as their motion was controlled by two opposing arcs (struck from pivots near the ends of the engine, and near its centre), they kept the cross-heads and piston-rods moving vertically up and down, as required.

The cylinders were provided with slide valves with a stroke of two inches, operated by two slip eccentrics fixed on the same axle; the rods which work them can be seen slanting up the sides of the boiler (Fig 23). The boiler itself was relatively large, with a diameter of 4ft 4in and a length of 11ft 6in; the flue tube was 2ft 1in in diameter. The piston-rods were 5ft 1in apart, which of course was also the length of the wheelbase. The wheels were of cast iron, of a form which had to be adopted owing to the limited capacity of the lathes at Shildon, with an inner disc 2ft 6in diameter and an outer ring of 3ft 11in, keyed with wooden wedges. The well-known photographs show the wheels as discs provided with 12 spokes projecting from the plane surface, and with circular holes in the outer rings to lighten the weight.

Locomotion made its trial trip at Newcastle on 11 September, 1825, and the opening day was fixed for the 27th of that month. The engine was drawn by horses to Heighington Lane, near Aycliffe, a few days later, and there unloaded. Just before the opening day the first passenger coach, the Experiment, was delivered at New Shildon; the locally built body was fitted onto an unsprung frame built at Robert Stephenson's works at Newcastle. A trial trip was made with this coach from Shildon to Darlington on the 26th, and all was ready for the grand ceremonies of the next day.

Long before dawn thousands of people began to converge on the railway, especially at Shildon, where the locomotive was awaiting its train; for the cable-worked inclines were also included in the demonstration. Ten loaded coal wagons were brought by horses from the Phoenix Colliery at Witton Park to the Etherley incline and worked by cable over the summit and down to the level stretch extending across the Gaunless valley near St Helens Auckland. A wagon-load of flour was attached, and the train of eleven wagons was drawn by horses to the foot of Brusselton incline; then by cable over the summit, and down to New Shildon. Here *Locomotion* stood waiting, attached to the passenger coach and twenty-one new coal wagons fitted with temporary seats. The trains were coupled together, making thirty-three vehicles in all. So far the proceedings had gone without a hitch.

The promoters however had not expected the enormous crowd which had collected, many of them determined on a ride. Three hundred seats had been reserved, but many more forced their way on to the train; the total number has been variously estimated at 400 to 600 people; the train measured 450 feet in length and must have weighed some 90 tons.

Trouble set in almost at once. After a few hundred yards a wagon derailed; when lifted on it immediately came off again, and had to be shunted off the track. The train went on to Simpasture, three miles from the start; but here a piece of oakum fouled one of the valves of *Locomotion*'s feed pump. Stephenson removed it and reported that all was now well; and the train completed the 8½ miles to the Darlington junction in 65 minutes, an average of 8mph, admittedly helped by a slight favouring grade.

At the junction a crowd of 12,000 people were waiting. Six of the coal wagons were shunted off for distribution to the poor of Darlington, and two others picked up. The eleven miles or so to Stockton were more difficult going, with some adverse grades; an average speed of 4 mph was maintained, but this perhaps includes a stop at Goosepool for water. Finally the train clanked over St John's crossing on to Stockton quay, to a tumultuous welcome from some 40,000 spectators. Arrival was 45 minutes late, but as 55 minutes had been lost by earlier delays, there was nothing to count against the engine.

The inevitable official banquet followed at the Town Hall, lasting till midnight, after which Stephenson must have retired to bed well satisfied with the day's work.

The triumphant opening day made a great impression on the public. But sadly, some years of frustration and confusion followed. *Locomotion* herself broke a wheel shortly after the opening day. *Hope*, the next in the class, arrived late and was found to be defective. Two of the *Locomotion* class exploded, no doubt because of careless handling. More troublesome, perhaps, were the difficulties which sprang from the railway's conception as a public highway. Coal traffic expanded rapidly; the single line soon proved inadequate; horse-drivers with their trains were obstinate and refused to give way as they should have done at the passing places; their wagons were ill-constructed, and the buffers and couplings did not match. Over some years the troubles were gradually resolved, the private traders were bought out and the wagons standardised. The line was doubled from Stockton to Brusselton by 1832, and its working began to resemble that of the much larger Liverpool & Manchester Railway, for which it had prepared the way. Stephenson at least was convinced that the new line must be a track reserved for the company's vehicles. Many landowners were enraged by the compulsory passage-way across their estates; their opposition had to be endured, by-passed, or bought off. Fortunately the enormous cost was soon repaid by heavy and profitable traffic.

8 – Paths of progress

Locomotive design made very little progress during the decade following 1816. The papers prepared for the Stockton & Darlington contract show that some new ideas were considered; but *Locomotion* as it turned out was a very conventional machine, with a straight-flue boiler, parallel motion, and the drive taken to unsprung wheels; though, as I have argued, the rough riding that resulted was offset by some improvement in power. But as the traffic on the railway increased, and horse-hauled trains were weeded out, the original four engines of the *Locomotion* class were not enough, even when all were at work. Extra power was clearly needed.

Engine 5 on the company's books was bought from R Wilson of Newcastle in November 1825. It proved a failure, though it was fitted with a return flue, a good feature lacking in *Locomotion*. It appears to have had four cylinders, and acquired the nickname *Chittaprat* from its unusually rapid puffing. It was soon laid aside, though parts of it were re-used for the much more successful *Royal George* of 1827.

No. 6, *Experiment*, was built in 1826–7 at Robert Stephenson's works in Newcastle, and was far more useful Its boiler measured 13ft by 4ft, as against *Locomotion's* 11ft 6in by 4ft 4in. Once again a single-flue boiler was used, though with ingenious devices to increase the heating surface. Water tubes were used for fire-bars, and the hot water under pressure was led forward through a pipe enclosed within the flue before returning to the boiler. Some of the exhaust was diverted to heat the feed-water, another useful improvement. But the most remarkable point was the layout of the cylinders and connecting rods, which clearly seem to have been designed to permit the use of springs, though there is no actual evidence that they were used on the engine in its original four-wheeled form. But this proved too heavy for the track, and the engine was certainly provided with a sprung frame when rebuilt as a six-wheeler.

The cylinders, 9in in diameter with a 2ft stroke, were mounted horizontally, and largely within the boiler, but projecting from its back-plate. Each of them drove a cranked axle, with another crank at its outer end driving a connecting rod; this passed down the side of the boiler with a gentle slope to drive the front pair of wheels. These were fixed to a rigid axle with the crank-pins arranged at 90° to one another; a good feature which Wilson, to his credit, had used on *Chittaprat*, and was adopted on all subsequent locomotives, as an extremely simple means of keeping the cylinders in phase without the complication of a return crank. It depended, of course, on using a single connecting rod for each cylinder, instead of a balanced pair driving on both sides.

A striking feature of the design was a chimney 16ft in height, with a sliding section which raised it another 8ft! This, one hopes, was used only for raising steam, when it might well have been helpful.

Experiment did a good deal of useful work, but Hackworth evidently disliked the single-flue boiler, and rebuilt her with a return flue in 1830.

The railway still needed more locomotives, and Hackworth set about building one at Shildon, using parts of the discarded *Chittaprat*. This was no mere rebuild, however; what emerged was the extremely powerful *Royal George* of 1827, in which several new features made their appearance.

The engine was a six-wheeler having a boiler 13ft by 4ft 4in, with the

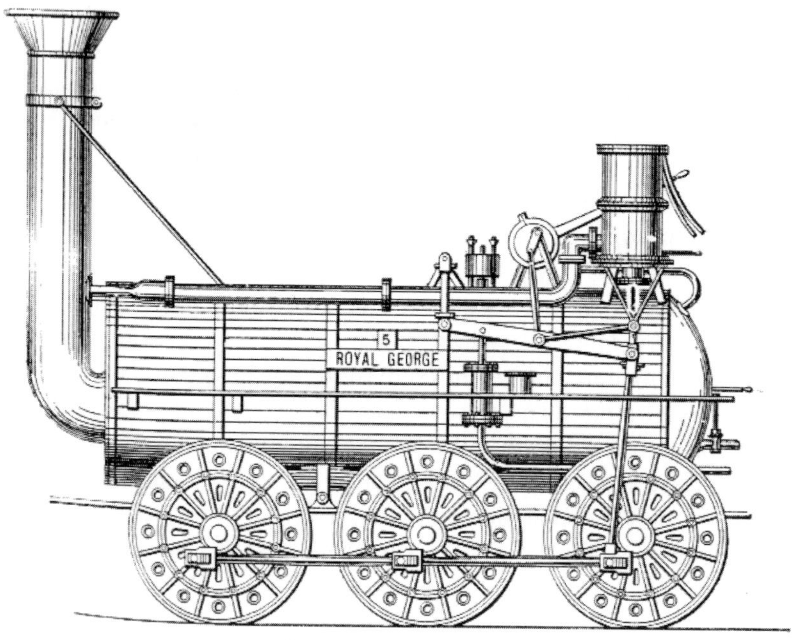

Fig 24 – Royal George*, 1827*

well-tried return flue. This combination provided a heating surface of 141 square feet, more than double that of *Locomotion*. Pressure was again 50lb per square inch. The cylinders were arranged, as on *Experiment*, to connect with opposite ends of the same axle, but with a much more elegant and straightforward layout. The cylinders, mounted beside the boiler, drove vertically downwards on to the rearmost coupled wheels, and were provided with a neat and compact form of parallel motion. On the debit side, the vertical drive meant that the driving wheels could not be sprung, though plate springs were provided for the other two coupled axles.

Hackworth's later locomotives of a similar design avoided the difficulty of unsprung wheels by connecting the cylinders to a dummy crankshaft, with horizontal connecting rods taking the drive to the three coupled axles, for which springs were provided. But this entailed mounting the cylinders well beyond the rear end of the boiler, which gave an uncomfortable lurching motion. Such was *Wilberforce* of 1831. A later development was to set the cylinders at an angle, like those on the *Rocket*. By 1842 they measured 14½in by 20in, with a working pressure of 70lb; but Hackworth persisted with the return-flue boiler and cumbrous double tender, which other designers had long abandoned. Progress took a different path, and though many different designs appeared from 1828 onwards, the two Stephensons were by far the most creative engineers.

Robert Stephenson returned from America in November 1827, after a perilous journey, and at once set about reorganising the Forth Street works, which had suffered neglect through his father's difficulties and embarrassments over the faulty initial survey for the Liverpool & Manchester Railway. Their first concern was a boiler with improved heating surface, without using the clumsy expedient of a return flue and additional tender. The new engine was ordered for the Liverpool & Manchester line, but was transferred by agreement to Bolton & Leigh, as this was nearer completion. It connected at Leigh with the Bridgewater Canal, and was subsequently extended southwards to a junction with the L&M.

The great innovation was to use two parallel flues. This was much simpler to construct than a return flue, though it was not easy to join the two flues together at the front end; at one time two chimneys were suggested! But given the same nominal heating surface, two flues would give better results, as the whole surface lay within a few feet of the fireplaces, whereas in a return flue of more than double that length the vapour would have given off a large part of its heat before reaching the far end; more economical, no doubt, but less effective.

The design took over from Hackworth the greatly simplified connection to the wheels from the cylinders, but with an important improvement; instead of driving vertically downwards, they were mounted in a sloping position, as in *Rocket*, within easy reach of the driver. This was frankly a compromise, but it was good enough to enable the driving wheels to be sprung. And with a much smaller boiler measuring 9ft by 4ft the total weight without the tender worked out at 7 tons, instead of 12, so that four wheels were enough. The nominal heating surface was much less than in the larger engine, 66 sq ft as against 141, but as I have argued, more efficient; the cylinders were 9in by 24in, as against 11in by 20in. The new engine used wheels made of wood with a wrought iron tyre; these were much less liable to breakage than the cast iron wheels used on the *Locomotion* class, and were repeated in several subsequent designs.

Another excellent innovation was a device for an early cut-off of the steam supply to the cylinders. This worked independently of the slide valves; a plug valve was mounted on a spindle which passed vertically upwards through the boiler and was driven by gearing from the back axle. An improved and greatly simplified version of it was supplied for an 0–6–0 engine of 1829, No. 7 of the S&DR, but it seems to have had no permanent effect on locomotive design; I can find no trace of it on the Rocket.

The new engine was an immediate success, and was given the name *Lancashire Witch* at the opening ceremony in August 1828. Besides a marked economy of fuel it had a fair turn of speed, attaining 11–12 mph running light. More important, of course, was its performance under load; designed to haul 20 tons at 7mph, it was actually recorded as pulling 37½ tons behind the tender at 8mph up a slight rising gradient of 1 in 440.

During 1828–9 a variety of designs for four- and six-wheeled locomotives was prepared by the Stephenson firm, not all of which were built. They all used the slantwise mounting of the cylinders, together with sprung wheels. A remarkable development of the twin-flue design was *Twin Sisters*, an 0–6–0 with two separate vertical boilers. The idea, no doubt, was to obtain a much larger heating surface than was possible with the much smaller and lighter *Lancashire Witch*; and the engine was in fact fairly powerful, propelling a load of 54 tons at 6–7mph, though the cylinders were no larger, at 9in by 24in. The design was not repeated, however; one might suppose that two medium-sized boilers would waste more heat than one larger one, and would cost more to maintain.

Two further Stephenson locomotives of which very little is certainly known were those sent to France at the order of Marc Séguin. Séguin had laid out a railway some 30 miles long from Lyon to St Étienne, close to the upper Loire, 'the only practicable means for realising the great benefit, so long desired, of a junction between the Loire and the Rhône'. It had easy inclines and curvature, no small matter considering the high ground to the west of the Rhône valley, and was intended for locomotive haulage. Séguin with other engineers visited England between December 1827 and February 1828, inspecting factories and railways and meeting George Stephenson. One of the engines was sent to Arras, the other taken to Lyon as a model for those which Séguin was to construct.

Séguin had some highly original ideas about boiler design; in fact he had applied for a patent for a multi-tube boiler in December 1827, just before his visit to England. The same idea had occurred, no doubt independently, to Henry Booth of the Liverpool & Manchester Railway; but Séguin's boiler was totally different from that introduced by Robert Stephenson on *Rocket*. He criticized the steam-raising capacity of the English locomotives, as received. These, then, most probably had the single-flue boiler which the Stephenson firm had used until *Lancashire Witch* came out in 1828.

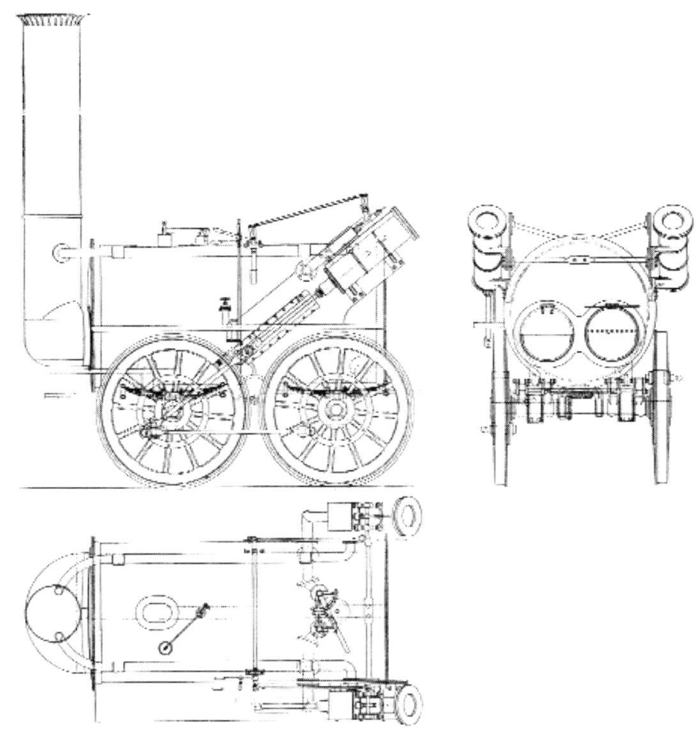

Fig 25 – Lancashire Witch, *1828*

On the other hand, the layout of the engine and motion as ultimately built by Séguin is quite different from anything recorded as having been built by Stephenson, and may well have been used, at Séguin's suggestion, on the two trial engines. The design was an 0–4–0 with cylinders mounted at a low level between the driving wheels. They drove vertically upwards on to the centre-point of long balance beams working connecting rods at both ends to serve the driving wheels, though the crank-pins were also connected by coupling rods. The piston rods were guided by an extremely elegant and correct form of parallel motion. The design enabled the shocks produced by irregularities in the rails to be evened out between the two cylinders, so that springs could be used. The weight of the engine, without its tender, was quite moderate, at a little under 6 tons.

Séguin's boiler, again, was quite unlike anything seen in England, but probably worked well. The boiler shell measured 3ft 2in by 11ft 8in, and carried a pressure of 60lb. A very large fireplace was provided, the grate measuring 2ft 7in by 4ft 9in, about 12sq ft, double that of *Rocket*, with a water jacket. Thence the hot gases went forward through a flue embracing the lower half of the boiler, and entered a casing which gave access to its front plate; they then passed through 43 tubes of 1½ in diameter (compared with the 25 3in tubes of *Rocket*) and were discharged through a short stumpy chimney at the rear end. The heating surface is given as 231sq ft, as against a little under 140 for *Rocket*; though in making these comparisons one should of course remember the extremely light weight which was required of the latter.

Initially the draught for the fire was provided by two rotary blowers in enormous wooden casings mounted on the tender and led forward to the engine by flexible leather hoses. The arrangement was mechanically sound, but suffered the disadvantage or relying on the engine's movement to supply the draught. This would then be fierce when the engine was running fast, when it might be gliding downhill under easy steam, and feeble when it was most needed, with the engine labouring slowly up against the grade. The great advantage of the blast-pipe was that the draught matched the output of power from the engine. Séguin described the difficulties presented by his original device; hence his 'happy discovery' of the use of exhaust steam. An early print of the engine at work shows no trace of the blowers.

In service the Séguin engine made a distinctive sound quite unfamiliar to English ears – as I can attest, having observed its replica at work at Lucerne in 1997. Stephenson's engines produced a distinctive puffing rhythm like their modern successors. On Séguin's machine only a smooth column of smoke unmixed with steam emerged from the chimney; the exhaust was discharged with a sharp hissing noise from each cylinder alternately. Thus useful energy was wasted, and further power was needed to drive the blowers.

Séguin has many claims to originality, and in 1828 his work was probably well in advance of the English engineers. On the other hand, he could afford to take his time, whereas Robert Stephenson was working to a tight time limit. In the end *Rocket* made her début just a fortnight before Séguin's engine took the road.

Despite their success these engines had little permanent effect on French practice. With the appearance of *Patentee* in 1834, English practice established a commanding lead, together with later variants, such as the outside-cylinder engine pioneered by Alexander Allan; not to mention the rear-wheel drive engines introduced by Crampton. The extremely popular and characteristic French design of outside-cylinder 2-4-2s, again, was a development of Robert Stephenson's long-boiler design, with the very desirable addition of a pair of trailing wheels to give the required stability and accommodate a larger firebox. From this point on, French designers went their own way, and only much later came to have a beneficent influence on English work.

9 – Rainhill

The level stretch of railway at Rainhill was the scene of the trials arranged by the Liverpool & Manchester Railway to decide whether to use locomotives or stationary engines hauling the trains by means of cables. The issue might have seemed straight-forward; granted that locomotives were not 100% reliable, it was easy to replace one by another in the event of failure; whereas a single breakdown of a stationary haulage engine would paralyse the whole system at once. Moreover, locomotives could be introduced in instalments as the traffic, and the income, built up. Nevertheless a competition was advertised, and severe conditions laid down; in the first place, a weight limit. The engines must not weigh more than 6 tons, if carried on six wheels, or 4½ tons if on four. They must consume their own smoke, and must draw a load of three times their own weight (or 20 tons for a six-wheeler) at 10mph with a boiler pressure not exceeding 50lb per square inch. The boiler was to be submitted to a test pressure of water at 150lb; and both engine and tender were to be supported on springs. The test track was 1½ miles long, and the locomotives had to reverse at each end and cover a total distance of 60 miles.

The rules, as revised, were remarkable, and allowed the engine's tender, if provided, to count towards the load to be hauled; thus the useful load would be well below the original demand. All in all, one may say that the engine's weight limit was severe, while the useful load to be hauled was very modest; in the case of *Novelty* under 7 tons, and 13–14 for the others.

Some very curious entries were submitted, including one worked by a horse; another, Burstall's *Perseverance*, was damaged in transit and could not be exhibited until the last day of the trials, after which Burstall withdrew it; it weighed under 3 tons, and was clearly underpowered. Thus only three locomotives had to be seriously considered; *Novelty*, entered by Messrs. Braithwaite and Ericsson, *Sanspareil* by Timothy Hackworth, and *Rocket* by George and Robert Stephenson.

Of these, *Novelty* was easily the smallest and lightest; her weight is not agreed; the official figure is as little as 2 tons 15 cwt, but it may have been around 3 tons empty, and something under four when charged with coal and water. This is the more remarkable in that she was, in modern terms, a tank engine, whereas her two main competitors were each equipped with a substantial tender whose weight did not count towards the limit of 4½ tons. Her design was unconventional, and included some excellent features; in particular she rode far more easily than the other entries, and attracted attention by her neat appearance, with an abundance of copper sheeting; and her flag was an imaginative touch, which helped to endear her to the crowd.

Her wooden frame was supported through springs on four light-weight wheels with rod spokes stepped in and out in the manner of modern bicycle wheels. Her two cylinders drove vertically upwards; this was a remarkably smooth-running design, since they were placed near the centre-line of the engine, and the drive was taken through a pair of intermediate levers, from which two further connecting rods carried it to a double crank axle for the driving wheels. Since these rods were nearly horizontal the driving wheels could move up and down on their springs without interfering with the motion of the pistons – a very desirable feature. The locomotive rode well and looked attractive, so that her favourable reception is easily understood.

She had four wheels 4ft 2in in diameter and was as entered, a single-driver, though the wheels could be coupled together 'if necessary'; the wheel-base was 6ft 6in. The cylinders were 6in x 12in; quite large enough for the very moderate requirements of the trial, though clearly too small for future service on the railway.

Her main disadvantage lay in the eccentric design of the boiler. The main portion of it was vertical in shape with a fire at the bottom surrounded by a water jacket. Draught was supplied from a pair of bellows driving compressed air into the enclosed space below the fire-box. The furnace continued upwards for a foot or two before narrowing to a vertical pipe passing through the steam and water space, and fitted with a lid or damper. This must have been kept open to serve as a flue when lighting up; it also provided a means of recharging the fire; a very questionable arrangement, as the fuel would arrive in a heap in the centre of the grate, whether or not it was needed there; moreover the driver had no access to the furnace, and consequently the fire could not be trimmed. Incidentally 'there were often singed eyebrows by neglecting to close the lower slide of the hopper before opening the top slide to insert fuel' (*The Engineer*, 1906, p121).

This vertical boiler was supplemented by a horizontal one, long and very narrow in shape, apparently only 13in in external diameter, which was traversed by a double-reverse flue much narrower than those normally used, through which the hot gases from the furnace were driven by the bellows.

When put on trial, *Novelty* suffered a series of crippling misfortunes, though our accounts are not agreed about their order and the dates on which they occurred. On the first day of the trials, 6 October, she made an impressive showing, attaining some 30mph unloaded, and was clearly expected to win; but there was a delay in preparing her for a second trial, which was put off till the following day; on her second appearance she exceeded 20mph with the prescribed load, admittedly a light one, before the trials had to be postponed owing to bad weather. On the third day, the conditions of the trial were altered, and it was agreed that *Novelty* should wait for a day before reappearing on 10 October; but this trial had to be given up owing to a driver's error, in closing a pipe between the feed pump and the boiler which should have been left open; though clearly the pipe itself should have been made capable of sustaining the full boiler pressure. Reappearing on the seventh day, she was once again disabled by the failure of joints in the seams of the boiler; and before the damage could be repaired, *Rocket* had fulfilled all the conditions without mishap, and was awarded the prize.

This summary relies on the report of the trials given in the *Mechanic's Magazine*. But an independent account given in a letter by John Dixon mentions another disaster, which he says occurred first, namely the bursting of the bellows which supplied the draught for the fire. Probably we should agree that this really happened, but amend the date so as not to contradict the very careful account given above. It may be one of the 'trifling casualties' mentioned by Nicholas Wood as occurring soon after *Novelty's* arrival; but a perfect agreement between the accounts is not to be had.

Further reflections on her design should concentrate on two points; the draught arrangements, and the boiler design. First, it is clear that a forced draught derived from the movement of the engine must suffer the same dis-

Fig 26 – Novelty

advantage as we have observed in Séguin's machine; the draft would be weakest when it was most needed. But whereas Séguin used rotary blowers of enormous size, *Novelty's* bellows must have been small and compact to fit into the space provided; very probably they were overstressed for the work they had to do.

As to the boiler itself, our accounts are not agreed on its effectiveness. We are told that steam could be raised in a surprisingly short time: 40 minutes, it is said, on one occasion. But we should like to know more about the methods involved. In the early stages, the driver must have relied almost entirely on the vertical part of the boiler; for until the engine began to move and the bellows were at work, there would be little inducement for the hot gases from the fire to traverse the long serpentine pipe and warm up its horizontal, section. Possibly, then, steam was said to have been raised when the engine was capable of moving; it would require further time before both parts

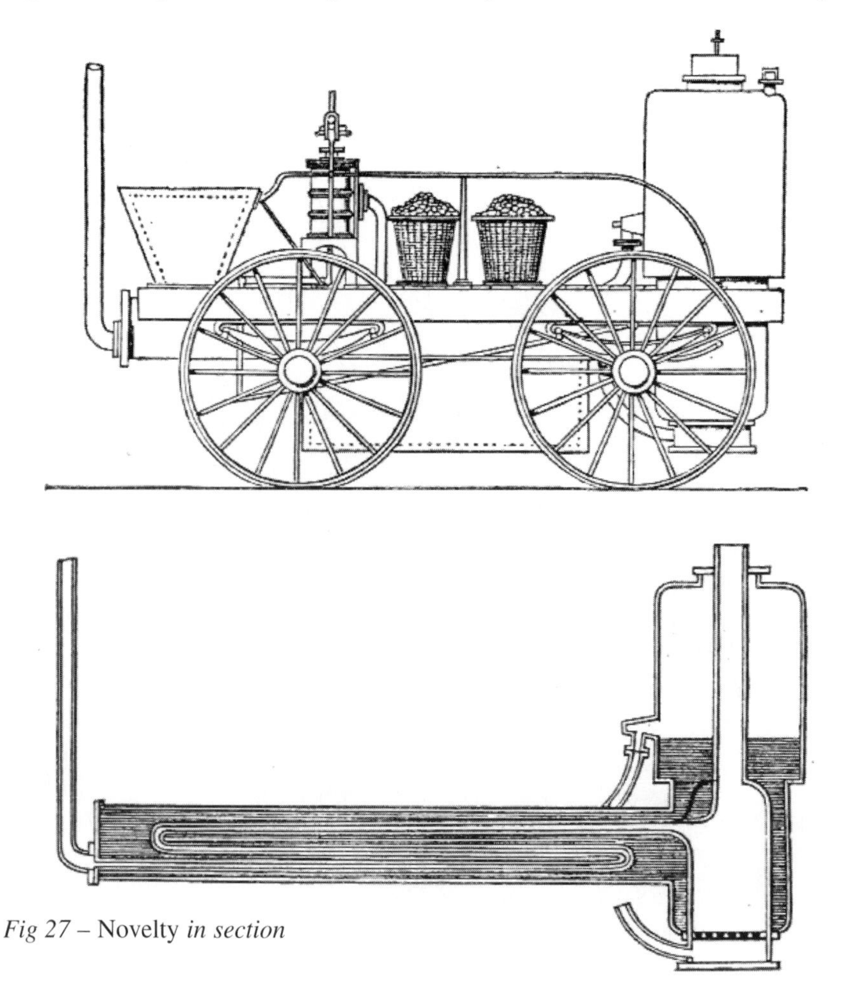

Fig 27 – Novelty *in section*

of the boiler became fully active and hard work could begin. After a mere 40 minutes, it may be that the water in the horizontal boiler was hardly more than lukewarm.

It is curious that *Novelty's* designers failed so entirely to appreciate the virtues of the exhaust steam blast. Obviously the trial locomotive was far too small and light to meet the demands of the new railway but when they came to present a pair of much larger ones working on the same principle, they replaced the bellows by a fan working at the exit from the vertical boiler. This was utterly ineffective, and the two locomotives were total failures.

Sanspareil, Timothy Hackworth's entry for the trials, was basically a cut-down version of his very successful *Royal George*, with the same arrangement of a return-flue boiler, which put the chimney alongside the fire-hole door, with the driver at the rear end in easy reach of the cylinders. This meant that the tender had to be pushed in front of the engine; though at the trials themselves this was no disadvantage, as all the locomotives had to traverse half the required distance in reverse. There was, however, at least one important difference from *Royal George*. Hackworth now abandoned the parallel motion, and reverted to slide bars to guide the piston-rods; he therefore had to give up the peculiar arrangement of mounting the eccentrics on a separate shaft as part of the motion. This was a very compact design, but meant that the connection of the eccentrics with the driving axle was indirect. On *Sanspareil* the eccentrics were mounted on the driving axle, and connected with the slide valves by long vertical rods curving up behind the domed end of the boiler.

Fig 28 – Sanspareil *(1829)*

Fig 29 – Sanspareil *(1829)*

The boiler was 4ft 2in in diameter, and 5ft long; but to increase the length of the flues, it was supplemented by an extension fitting fairly closely round the fire-door and a section of the return flue next the chimney, with a small water space enclosing them. The driving wheels measured 4ft 6in, and were necessarily placed close together, giving a wheelbase of no more than 4ft 8¼ in The cylinders were vertical, as in *Royal George*, and both pairs of wheels were unsprung, contravening a requirement for the trial.

Sanspareil was well over the weight limit, at 4 tons 15½ cwt, as against the 4½ tons allowed; but the judges allowed Hackworth to give a demonstration. It appeared to make plenty of steam, but at the cost of a prodigious consumption of fuel, much of which was thrown out though the chimney as a rain of red-hot cinders. Hackworth rightly claimed the credit for bringing the exhaust steam together in a blast pipe of the modern type, as against two separate exhausts; but the effect was far too fierce. It seems a pity that no one thought of providing a by-pass valve to release part of the exhaust to the air, and put the blast to that extent under the driver's control.

Sanspareil made a promising start, and had covered about half of the required distance when a leak developed in the feed pipe to the boiler, and despite all Hackworth's efforts the water level fell so low as to melt the fusible plug over the fireplace, bringing the trial to a precipitate end. (Some accounts mention a further defect, namely a cracked cylinder, but the evidence for this is not conclusive. Hackworth, sadly disappointed, asked to be allowed a second demonstration; but this was refused; the engine in any case had not fulfilled the conditions laid down.)

Hackworth's many friends were aggrieved at the decision in favour of *Rocket*; they argued that *Sanspareil* was nevertheless a powerful machine; as indeed it was, though ungainly in appearance and awkward in action; John Dixon gives a vivid account of the impression it made: 'he rolls about like an Empty Beer Butt on a rough Pavement'. But their case was soon disproved; the railway bought the engine, probably as a favour to Hackworth, and to disarm suspicion. But it proved unsuccessful in service, as well as outrageously uneconomical, and was soon discarded. The main reason, however, was, its awkward riding at high speed. It was sold in 1832 for £110 and worked adequately on heavy goods trains on the Bolton & Leigh railway until 1844.

A much better case for Hackworth can be made by considering his difficulties; he had to maintain a stock of ailing locomotives, as well as desiging their replacements, for the quite different conditions required at Shildon; he could not, as Robert Stephenson clearly did, give his full attention to his entry at Rainhill. In my own opinion, his principal mistake lay in designing a boiler with the conventional diameter of 4ft 2in. Since its length could hardly be reduced, this alone ensured that the engine would be too heavy to meet the conditions. He could have avoided this by reducing its diameter, say from 50 inches to 40, or 3ft 4in, as on *Rocket*. This would have reduced its weight roughly in the proportion of 25:16, or some 36%, and would have allowed its length to be increased, say by a foot, or 20%, while remaining well within the weight limit; the extra length would have allowed a very desirable increase in the length of the flues, and made it possible to extend the wheelbase, giving far better riding. Again, the question of springs could have been reconsidered. Hackworth was compelled to omit them, partly because of the extra weight involved, partly because with his very short wheelbase they would have produced an unacceptable bouncing movement at speed. Given a longer wheelbase, they might well have been adopted. Indeed it looks as if Hackworth himself had at one time considered the possibility, since he actually used cylinders 7in in diameter by 18in stroke, longer and narrower than those of *Rocket*, while giving the cylinders an internal length of no less than 24in. The extra length would allow for the irregular up-and-down movement of the coupled wheels, and the small diameter meant that the extra clearance required at the two ends would not involve too great a waste of steam.

In any case, the fireplace was almost certainly made too large, with firebars no less than 5ft long. As a result, the engine's capacity for generating heat far exceeded its ability to transmit the heat to the water. Thus the temperature of the exhaust gases was too high, and the consumption of fuel enormous; the effect was of course made worse by the undesirably fierce blast. These observations, however, were often ignored at the time of the trial, though the ejection of red-hot coals must of course have been noted.

Fig 30 – Rocket, *original form*

Fig 31 – Rocket *in section*

Rocket's victory was so complete and overwhelming that it sometimes appears as a walk-over brought about by a stroke of intuitive genius. But this is to overlook the very careful planning of the design, and the hard work involved in freeing it from the inevitable defects that appeared in the course of manufacture. The Stephensons were clearly well aware of the time limit, and allowed sufficient margin for this to be done.

Rocket's main features are quickly stated. In many ways she resembled *Lancashire Witch*, with slanting cylinders mounted over the rear wheels and driving forwards, but she was built as a single-driver; Stephenson had shrewdly reckoned that the weight could be reduced by using only a small pair of carrying wheels and omitting the coupling rods. The driving wheels were 4ft 8½ in, the cylinders as aforesaid, 8in by 17in, and the boiler 3ft 4in in diameter and 6ft long.

The most important innovation, of course, was the multi-tube boiler. This apparently was not Stephenson's own unaided inspiration; it was suggested by Henry Booth the treasurer of the railway and, as we have seen, Marc Séguin was already at work on a quite different design in 1829. In principle it was an extension of the idea already adopted in *Lancashire Witch*; but it involved a break-away from the traditional flue, which both contained the fire-place and traversed the boiler, in favour of a separate fire-box; and here the design adopted was the prototype of all subsequent locomotives, with only the small modification of enclosing the fire-box in an extension of the boiler shell. *Rocket*'s fire-box was a separate construction communicating with the boiler proper by three pipes; though at this early stage this may have brought some advantage in enabling the two elements to be constructed and tested separately, and in fact the fire-box was made by another firm. It was surrounded by a water space, and its upper portion overlapped the rear end of the boiler proper, so that the hot gases could pass through 25 fire-tubes. This relatively complicated design led to some difficulties in manufacture; in particular the boiler end-plates were found to have been made too light, and a number of extra stays had to be introduced to overcome their tendency to bulge under the pressure of water at the required 150lbs.

At the trials themselves Stephenson was content to run *Rocket* under fairly easy steam, reeling off lap after lap at 14 to 15mph as against the 10mph required by the rules. But on the last eastward trip she was given her head, and traversed her distance in a fraction over 29mph. Two other striking performances were recorded; on one occasion a load of 40 tons was drawn at the 14mph achieved with a much lighter load at the trials, Moreover, a load of 18 tons was drawn up a gradient of 1 in 96 at 8mph.

Rocket was, in many respects, the prototype of all subsequent locomotives, although it had a short life in its original form. It was of course deliberately made very small and light to satisfy the requirements of the trial, and in the next two years it was superseded by larger machines built to the same general pattern, but with a progressively increasing number of fire-tubes and the important improvement of a smoke-box, which provided the boiler with a 'front door', which made it much easier to clean the fire-tubes and also to remove the mass of ash and clinker which was bound to accumulate below the chimney; while much better riding was obtained by mounting the cylinders in a horizontal position; and *Rocket* herself was soon rebuilt to incorporate these improvements. However, on Stephenson's advice she was withdrawn from first-line duties and put to lighter work, hauling ballast trains and the like. She remained in the Railway's service until October 1836, when she was sold for £300 to a Mr James Thompson of Carlisle. Contemporary prints, it is true, show *Rocket* in her unrebuilt form attached to quite substantial trains; but no doubt this is merely artistic licence. At the opening of the railway on 15 September 1830, the larger and more powerful *Northumbrian*, driven by George Stephenson himself, was assigned to the principal train. *Rocket*, however, was given an honourable place in the procession.

10 – Developments in England

After her resounding success at Rainhill, it may seem strange that *Rocket* was almost immediately regarded as obsolete and relegated to secondary duties. But the reasons are clear and comprehensible. First, she was deliberately designed to meet the conditions imposed at the trial, which demanded a very severe limitation of weight and required only a very modest tractive effort. As it turned out, she was powerful enough to meet these conditions with ease, hauling an appreciable load up inclines designed with the possibility of rope haulage. Nevertheless the rapid growth of traffic soon called for more powerful machines, and for the first year of operations these were enlarged and improved versions of the same design. Second, the new features of that design opened the way to improvements which would not have been possible otherwise; above all the increase in the number of boiler tubes, which greatly enlarged the heating surface and opened the way to larger cylinders, but also the tidying up of the boiler, with a smokebox at the front end and the incorporation of the fire-box in the boiler shell, which avoided the waste of heat in *Rocket's* separate firebox and the tubes connecting it with the boiler.

No time was lost. *Rocket* was almost immediately followed by the *Meteor* class of four engines, ordered in October 1829, whose boiler included 88 x 2in tubes, as against 25 x 3in, and made it possible to use cylinders 10in x 16in, as against 8in x 17in; the cylinders were lowered to a nearly horizontal position, a modification which was soon made to *Rocket* herself, and the total weight went up to 5 tons 4 cwt. The *Phoenix* class of two engines followed, ordered in February 1830 and delivered in June. These had the boiler lengthened to 6ft 6in and provided with a smokebox, and 11in cylinders. Next came *Northumbrian*, delivered in August 1830, and the *Majestic*, January 1831; here the heating surface was further increased by using 132 tubes 1⁵/₈ in diameter and using a longer firebox, though 11in cylinders were retained. These engines also made extensive use of vertical plate frames, which later became a standard form of construction. With these modifications the weight increased to 7 tons 7 cwt.

The smokebox replaced the detachable plate at the base of *Rocket's* chimney; it gave easy access to the fire-tubes for cleaning, and allowed for the removal of the ash and clinkers which collected in front of them.

As a deviant from this regular succession there appeared *Invicta*, the single engine constructed for the Canterbury & Whitstable Railway, whose function was to connect Canterbury with the steamers plying between Whitstable and London. It was a 0–4–0 with slanting cylinders mounted at the leading end, otherwise a little like *Lancashire Witch*, but with a *Rocket*-

type boiler. This was unwisely removed and replaced by an old-fashioned straight-flue boiler. But the engine's opportunities were limited, as much of the line was steeply graded, and intended for cable haulage, with the exception of a mile or so at the Whitstable end.

In the mean time Robert Stephenson had been giving careful thought to the design of further locomotives. Despite its manifest improvement on its predecessors *Northumbrian* did not ride really well, and Stephenson rightly thought that much easier travel could be achieved by mounting the cylinders close to the engine's centre line; moreover if they were placed beneath the smoke box the waste of heat could be reduced by keeping them close together and eliminating exposed connecting pipes. It would, of course, be necessary to use a double-cranked axle; but there was precedent for this in the design of London road carriages. In some other ways, notably the use of outside wooden frames, *Novelty* may have acted as a model.

Planet was an outstanding success, and many engines of this type were constructed of the same basic design, but with slight variations. As we shall see, quite a number were exported to the United States, and a close copy was built by Baldwins in 1831. Moreover an enlarged version required for heavy duty on the inclined planes appeared as a 0–4–0 in *Samson*, delivered early in 1831, followed by *Goliath*. All the main dimensions were increased, with a boiler 7ft x 3ft 6in containing 140 tubes of $1^{5}/_{8}$in diameter in *Samson* and 132 in *Goliath*; the grate area was increased from $6^{1}/_{2}$ to $7^{1}/_{2}$ square feet, and the cylinders were 14in in diameter; at the same time the diameter of the driving wheels was reduced to 4ft 6in. The weight of these engines was 10 tons, all of it of course available for adhesion.

We have so far considered the locomotives built for the L&MR by Robert Stephenson; but other designers and builders were soon at work, of whom the most talented was probably Edward Bury. Bury favoured a boiler with a D-shaped firegrate,

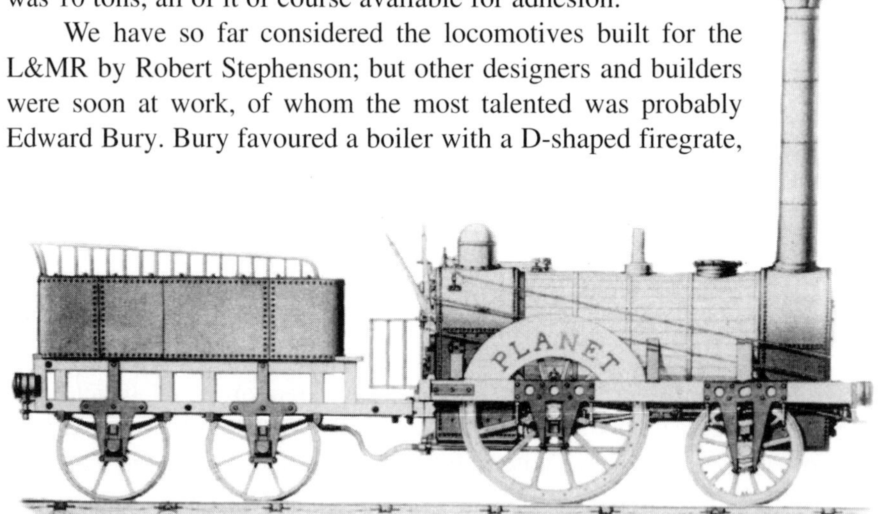

Fig 32 – Planet, *1830*

and the firebox itself of circular form surmounted by a dome; but this form of firegrate was not easily enlarged, whereas a square or oblong one could simply be lengthened. He also used driving wheels 6ft in diameter, while Stephenson held that 5ft maximum was desirable. In 1831 Bury submitted one of his engines – *Liverpool* – for trial; it was rejected for various reasons, including an inability to make steam. But by far the most important point was an accident brought about by a broken crank axle.

Bury favoured simple inside frames. But in this case any fracture of the crank axle would lead to the immediate collapse of the unsupported driving wheels, possibly with disastrous consequences. To avoid this Bury coupled the engine and tender tightly together. Stephenson preferred outside frames; but his main objective was to provide additional bearings between the driving wheels and the vulnerable crank, that the wheels were always located between two supports. In normal service, however, the whole weight of the engine was borne on the outside sprung bearings. It seems that some attempt was made to provide springs for the inside bearings also; this was soon given up, no doubt as an avoidable complication.

The one really successful engine built by Bury for the railway was *Liver*; at Stephenson's insistence it was built with outside frames. It seems to have some improvement in performance on the normal *Planet* type. However as loads increased Bury still insisted that small four-wheeled locomotives could do all that was needed; this later caused serious trouble on the London & Birmingham Railway, for which Bury had contracted to supply the power. Here the use of three engines was by no means unusual, and on one occasion it is recorded that no fewer than seven were needed to get a heavy train going in bad weather.

Robert Stephenson, however, had soon realised that the increasing loads called for six-wheeled engines. He recommended their adoption in a report dated 5 November 1832; a month earlier he had already got out drawings for a 0–4–2 engine, which found a home on the Leicester & Swannington Railway, the first component of what became the Midland. The definitive design, however, was that shown in a drawing made for the design for a 2–2–2 patented in October 1833. Stephenson claimed that one of its main advantages was the use of unflanged wheels on the driving wheels, the leading and trailing wheels providing the necessary guidance, and relieving the crank axle of sideways pressure. At the same time, it became possible to use a very much larger boiler and firebox without imposing too much weight on any of the axles. A locomotive of this design was offered to the L&MR in April 1834; it was accepted and given the name *Patentee* .

Patentee's boiler was 3ft 6in in diameter and 7ft long; it contained 106 tubes 1 5/8in in diameter giving a heating surface of about 364sq ft. This was actually rather less than some of the *Planets*; but the real advantage of the new design lay in its much larger firegrate, nearly 10 sq ft in area. This combination provided for cylinders 11ft x 18in, a rather conservative improvement on the previous 11ft x 16in. The weight of the engine was 11 tons 9 cwt, but rather surprisingly it was the leading wheels that carried the heaviest weight – the cylinders no doubt contributing – whereas the weight on the drivers was only 4 tons 6 cwt, less than that of the *Planets*, at eight tons in all with rather more than five on the drivers.

The success of *Patentee* led to a rapid development of six-wheeled engines of various types. We have already mentioned the very early use of an 0–4–2 design. On the whole the 2–2–2 was preferred for passenger trains, but the 2–4–0 was an obvious alternative; for goods trains the choice lay between the 0–4–2, the 2–4–0 and the 0–6–0.

Before long engines of all these types, with the characteristic outside frames, found their way overseas. The first orders apparently came from Belgium in 1834. The largest of the first three was *L'Éléphant,* built by Tayleur & Co as a 2–4–0, while two 2–2–2s were delivered by Stephenson; all three took part in the opening of the Brussels–Mechlin line on 6 May 1835. Many further orders followed, and both George and Robert Stephenson received decorations from King Leopold.

Germany came next, the project for the Nuremberg–Fürth Railway had already been published in 1833 but, owing to a disagreement about the price, the first locomotive, *Der Adler,* was not delivered till 1835. It was a smaller

Fig 33 – Patentee *(1833) in section*

edition of the *Patentee*, with an overall weight of only 6 tons 12 cwt. Similar engines were delivered to the Berlin–Potsdam line and to other German railways. In Russia the introduction of railways was warmly supported by Grand Duke Nicholas, and three locomotives were ordered from Timothy Hackworth, Robert Stephenson and Tayleur & Co. Extraordinary claims were made for the speeds reached by these machines, most probably running light; the Stephenson 2–2–2 is said to have run at 65½mph, and Hackworth's at 72!

The most notable of the Stephenson engines supplied to France was *La Victorieuse*, built in 1837 for the Versailles Railway as a 2–4–0. In direct contrast to the German precedent, this locomotive was abnormally large and powerful. It weighed about 13 tons, having cylinders 15in x 18in, a grate area of about 11¼square feet and a heating surface of 595sq ft. The driving wheels were relatively small, at 4ft 6in, so that the output of power must have been very large for this early date.

However the growth in locomotive power is perhaps best seen in the 0–6–0s designed and built by Stephenson for the Leicester & Swannington Railway, a fairly short line carrying a heavy coal traffic. As early as 1833 drawings were got out for a locomotive with a total heating surface of 656sq ft, grate area 10sq ft and cylinders 16in x 20in, again with 4ft 6in wheels. This engine, named *Atlas*, went into service the next year. It was tried out again Edward Bury's four-coupled Liverpool, after Bury had rashly boasted that 'Whatever Stephenson's engine could do, his could do'. *Liverpool*, having only 12in cylinders against its rival's 16in, was soundly defeated; its cylinders, of course, were smaller than its rival's in the ratio of 12:16, or 3:4. All the same, Bury obstinately persisted that four-wheeled engines could do everything that was required, with results that we have seen.

It seems, however, that British designers were rather slow to introduce engines for goods service much larger than *Atlas* of 1833 and the rather similar *Hector* of 1839. The first example I can trace is the class of six engines built by Kitsons in 1848 for the Leeds & Thirsk Railway which had cylinders 17in x 24in, 1001sq ft total heating surface and 4ft 9in wheels. But this was soon exceeded by the very large 2–2–2 built by McConnell at Wolverton for passenger service in 1849 which had cylinders 18in x 21in and no less than 1539sq ft of heating surface. The reluctance to use much larger goods engines is perhaps to be explained by the conditions of service, especially the difficulty of operating very long trains due to the inadequacy of sidings, which was certainly a restraining factor at a much later date.

11 – Beginnings in America

American engineers' attention was naturally caught by the development of railways in England. The success of the Stockton & Darlington Railway, crude as it was, came in for notice; the much more important project of the Liverpool & Manchester aroused lively interest. In 1828 American engineers, as well as Marc Séguin, visited the works at Newcastle; two locomotives generally similar to *Lancashire Witch* were sent to America later that year. However, the first actually put to work was *Stourbridge Lion*, built in 1829 by the firm of Foster and Rastrick for the Delaware & Hudson Canal Company. Driven by Horatio Allen it made a venturesome trial run, including the timber trestle bridge over the Lackawaxen Creek, but with a weight of 7 tons it was thought too heavy for continued use.

It must be confessed that *Stourbridge Lion* was not a very distinguished representative of English practice, with its single-flue boiler and cylinders driving vertically upwards on to half-beams. But the much more successful *Witch* still turned the scale at 7 tons exactly. English engineers had learnt to construct their tracks to suit this weight. But in America conditions were very different; timber was abundant, whereas iron was in short supply. The Delaware tracks at this first trial used timber longitudinals with iron straps to protect them from wear; these were soon replaced by cast-iron fish-bellied rails in short lengths resting on wooden sleepers; but suspicion persisted, and *Lion* had to be laid aside.

Fig 34 – Stourbridge Lion

Unfortunately the use of timber rails with iron straps became fairly widespread, in the attempt at a quick expansion of railway mileage, and as speeds increased it led to unpleasant accidents. The straps were liable to fracture and the broken ends to bend upwards so as to poke themselves through the carriage floors and skewer any luckless passenger who was in their way. In the end these 'snakes' heads' were eliminated as all-iron rails became standard; but the change was slow in coming.

But if the quality of early American railway construction was poor, its quantity was most impressive. Railway promotion in England at first was a slow and expensive business; parliamentary approval had to be obtained, and large landowners placated. The first really important new main lines after 1830 were the Grand Junction, linking Birmingham with The Liverpool & Manchester, opened in 1837, and the London & Birmingham in 1838. In America no time was lost and construction began from a number of centres. An early success was the Charleston & Hamburg RR, chartered in 1828, with its first section completed in 1830, and opened throughout to Hamburg, 105 miles, in 1833. We shall also notice three railways further north on the Eastern seaboard which were prominent in the early years.

The Baltimore & Ohio was an ambitious project, designed from the first to link Baltimore with the important Ohio waterway. It opened with a double-track line from Baltimore to Ellicott's Mills, 12 miles, where construction was temporarily halted by a dispute with the canal company over a right of way; but construction was soon resumed, and the line was open to Frederick, 60 miles, in December 1831. The B&O ultimately reached the Ohio at Wheeling, 379 miles, in 1833; in the same year the Pennsylvania RR reached the much more important centre of Pittsburg from Philadelphia. Both routes involved laborious climbs over the Allegheny Mountains. The Camden & Amboy RR was the first attempt to provide railway communication between Philadelphia and New York. From Camden, across the river from Philadelphia, it ran north-eastwards for some 60 miles to the little harbour on the Bay of Amboy, from which New York could be reached by ferry, about 25 miles. The line naturally lost its importance when a through all-railway route was opened, and has disappeared from modern maps.

The beginnings of the Pennsylvania RR seem to be traceable to the charter granted in 1828 for a railway from Philadelphia westwards to Columbia on the Susquehanna River, 81 miles, opened in 1834. Columbia was later bypassed by routes which reached the river at Harrisburg, about 100 miles, one of them using the Philadelphia & Reading, chartered in 1833. Meanwhile the Philadelphia, Germantown & Norristown RR, whose depot is illustrated in an attractive early engraving, seems to have been of purely local importance. The engraving shows an unusual type of four-wheeled coach; its body is

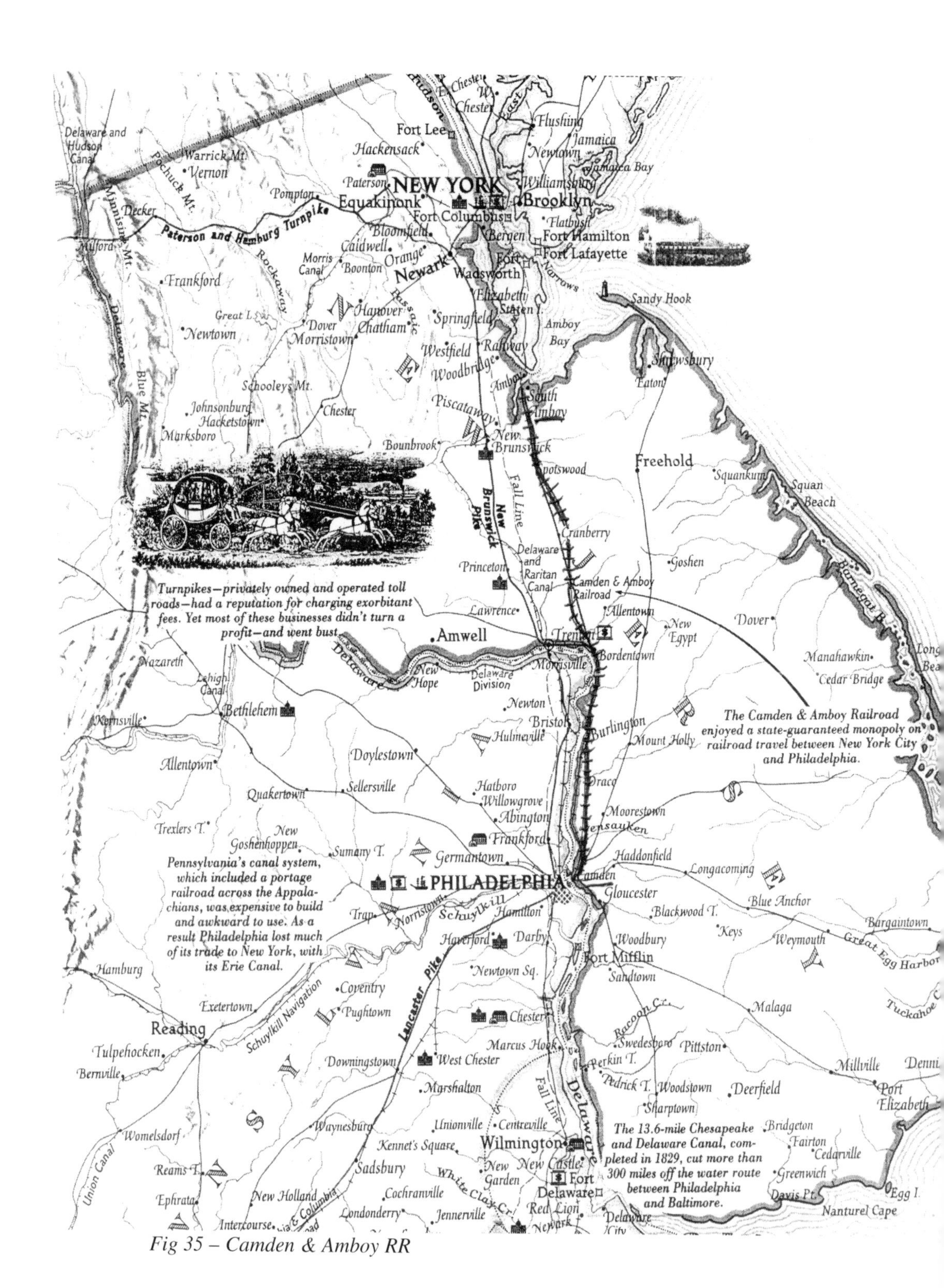

Fig 35 – Camden & Amboy RR

ornate and traditional, and apparently carried 20 passengers or so in a single large coach-style compartment; it is unsprung, but each one has its own brakesman, and the roof space is elaborately railed in, no doubt for the placing of passengers' luggage. The locomotive, a 2–2–0 of the *Planet* type, is of interest as being the first to be constructed at Baldwin's works, which had opened at Philadelphia two years before.

The engineer who best appreciated the limitations which governed his work was Peter Cooper, who had constructed an experimental engine, *Tom Thumb*, weighing less than a ton, for the timber-and-strap road of the Baltimore & Ohio Railway. It was tried out in August 1830, and was a simple four-wheeled truck carrying a vertical boiler and driving the wheels through gearing. A well-publicized race between the engine and a horse caught people's imagination. *Tom Thumb* with its car-load of directors started in good style, but was disabled by the failure of the bellows which supplied the draught, and the horse-drawn car came in first. Nevertheless the possibility of steam traction had been sufficiently shown to convince the doubters.

Fig 36 – Tom Thumb

Fig 37 – Best Friend of Charleston *(1831)*

The first full-size locomotive built in America to give useful service was the four-ton *Best Friend of Charleston*, introduced on the South Carolina Railroad (as it became) in January 1831. This once again had a vertical boiler, rather oddly perched ahead of the leading wheels, and slightly slanting cylinders, using normal coupling rods to connect the driving wheels. It could haul 50 passengers in six cars at a top speed of 21mph, and reached 30mph running light. Unfortunately its working life was short; it was destroyed by a boiler explosion in June of that year owing to the stupidity of the fireman who, annoyed by the roar of escaping steam held down the safety valve.

Vertical-boiler locomotives were also in use on the Newcastle & Frenchtown road, and on the Baltimore & Ohio, where a very influential machine, *Atlantic*, was put to work in 1832. This type held a monopoly on this important main line, as it became, until 1837, when horizontal boilers were introduced. Vertical-boiler locomotives became quite widely used in the USA, but were never very common in England; there is one lone survivor, *Chaloner*, still at work on the narrow-gauge Leighton Buzzard Railway.

Atlantic still survives in working order, pulling replicas of the original cars, four-wheelers of the traditional coach design, but open-sided except for the doors, and surmounted by an upper deck for passengers, with canopy-style roofs. One wonders whether these could have possibly escaped burning by sparks from the engine!

A much closer approximation to normal English practice was *De Witt Clinton*, the second American-built locomotive, which was put to work on the Mohawk & Hudson Railroad, one of the constituents of the later New York Central, in December 1831. It was a 0–4–0 with a rather small boiler sur-

Fig 38 – Atlantic *(1832)*

Fig 39 – De Witt Clinton *(1831)*

Fig 40 – John Bull

mounted by a prodigious dome and the chimney projecting forward of its leading end. The coupling rods were braced by wires passing over short king-pins above and below their central point. The well-known drawing shows it with a tender carrying the water-tank above the centre-line of the boiler, and a set of passenger carriages of extremely primitive design, with a single traditional-style coach body mounted on each four-wheeled underframe. These cannot have lasted long; a much more sophisticated three-compartment coach was already in use in the same year 1831 on the Camden & Amboy Railroad. *De Witt Clinton* made her maiden run from Albany to Schenectady, 14 miles, in 46 minutes, about 13mph; not quite as good as *Best Friend*!

The Camden & Amboy was also one of the first to import English locomotives of the *Planet* type, together with the very similar *John Bull*, an 0–4–0 with a circular firebox, put to work in 1832. It was criticized for its unsteady riding, and a cure was found in a rigid two-wheel truck which extended the wheelbase by six feet or more, and which became the prototype of the widely used American cow-catcher. The tender was provided with a roof and a veranda projecting over the footplate; this was later replaced by the standard American square cab and a bogie tender. In this form the locomotive remained at work for a number of years. It was brought out again for a celebration in April 1893, and with a two-coach train completed the 920-mile run from New York to Chicago without needing assistance. Indeed this same locomotive was steamed once more for its 150th birthday celebrations, and made several runs.

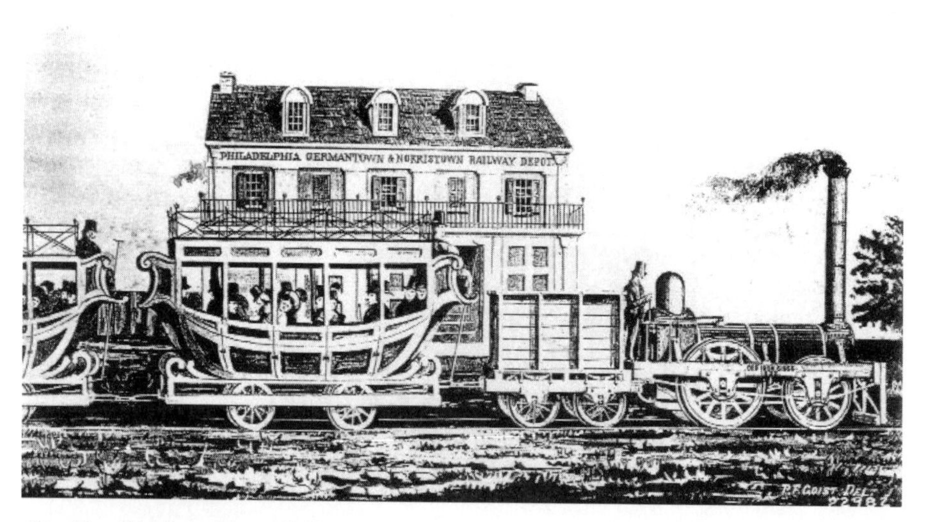

Fig 41 – Old Ironsides *(1832)*

An 0–4–0 with Edward Bury's characteristic iron bar frames was exported to America in 1833, and was named *Liverpool*. Before long American designers had discarded the wooden frames used earlier, and adopted bar frames themselves.

A considerable number of *Planet*-type locomotives, both 2–2–0s and 0–4–0s, found their way to America from Robert Stephenson's works between 1831 and 1836. But I cannot discover that any of the *Patentee* type, so immediately successful in England, ever crossed the Atlantic. The reason was probably the early adoption of the leading bogie; the *Patentees* were perfectly at home on the well-laid English permanent way; but Robert Stephenson himself had recommended the use of the bogie on sharply-curved American tracks to the deputation of American engineers as early as 1828. He was told that the B&O road intended to use curves with a radius of only 400ft, or six chains; such curves were used in England only at the turn-out lines at a few junctions, where a severe limitation of speed was enforced.

The first bogie 4–2–0 was *Experiment* built by John B Jervis for the Mohawk & Hudson Railroad in 1832. It had outside frames both for the engine itself and for the bogie; and similar construction was used on the first 4-4-0 built by Henry R Campbell for the Philadelphia–Norristown line in 1836–7. Similar outside-framed 4–2–0s were built by Stephenson for the

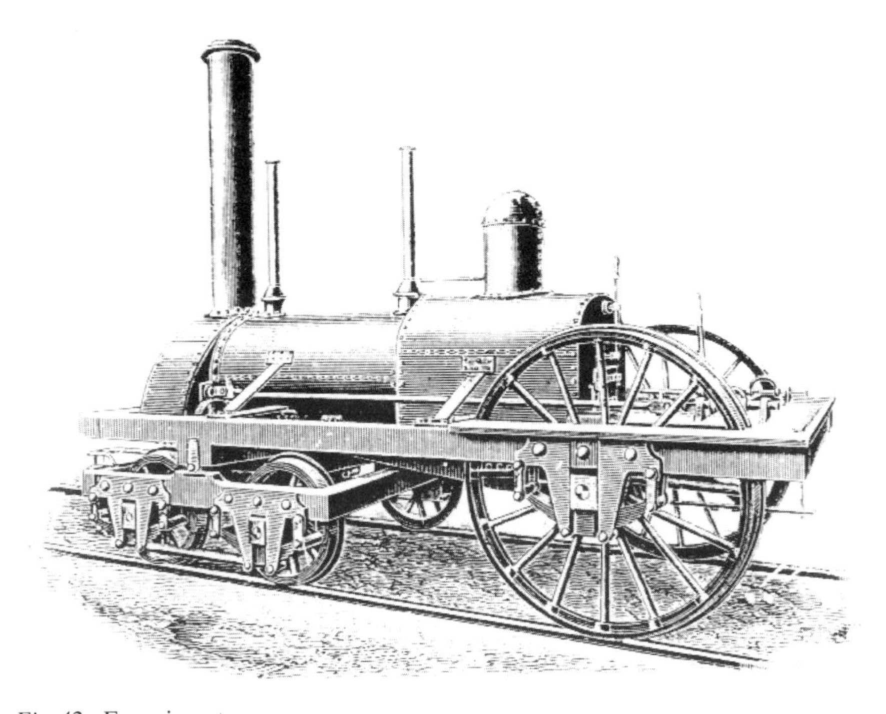

Fig 42– Experiment

American market in 1833–4; but both outside frames and inside cylinders were soon given up; the Norris 4–2–0s imported by the English Birmingham & Gloucester Railway in 1838 or soon after had inside frames and outside cylinders with a firebox surmounted by a domed top. However such engines soon gave place to the very widely used 'American' type of 4–4–0, with inside frames and outside cylinders, the valves being worked through rocking levers from eccentrics placed between the frames. It was probably a great advantage that American engineers, with few exceptions, concentrated on the development of this single type. In England the position was far more complicated. Stephenson himself had given up his distrust of crank axles at the latest by 1841, and was using inside frames and cylinders; the increased length of six-wheeled engines also made it possible to use outside cylinders with reasonably easy riding. Nevertheless outside frames combined with inside cylinders continued in use in England for a surprisingly long time, constructed at first of wood and then of the 'sandwich' type, where a wooden member enclosed between iron plates gave the frames a certain flexibility; and later still all of metal. Large outside-framed 2–2–2s were still in use on the Great Western Railway in the 1890s, when a derailment led to their rebuilding as bogie 4–2–2s, the best-known being the renowned *Duke of Cornwall* which took over the record-breaking train of 1904 from *City of*

Fig 43 – Outside-framed American 4-4-0

Fig 44 – Typical American 4-4-0

Fig 45 – Pioneer *(1836)*

Truro and averaged 80mph over most of the route between Bristol and London. The *Cities* themselves were also outside-framed. The Midland Railway used outside-framed 0–6–0s and 2–4–0s designed by Kirtley in the 1860s and 70s, many of which had an exceedingly long life; a 2–4–0 built in 1868 was not withdrawn till 1936, and one of the smaller-wheeled variety, still at work in 1945, was set aside for preservation; and Johnson, also on the Midland, used outside frames for his notable 4–2–2s, some of which remained at work into the 1920s. A still more surprising survival was the reintroduction of outside frames in the 1930s on the Great Western Railway, which needed some 4-4-0s for secondary duties on lightly laid lines, and used the well-tried parallel boilers from the *Duke* class married to the robust frames salvaged from the powerful mixed traffic *Bulldogs*.

In England in the 19th century there was no agreement about the best design, and some unusual combinations were adopted. Thus the very successful *Jenny Lind* design of 1847 used inside frames for the driving wheels, but outside frames for the other two pairs, at first sight a curious idea. But it has its advantages; weight can be saved by abandoning outside frames on the driving wheel, if the crank axle can be trusted; whereas outside-framed leading wheels give a better balance, as the springs can be placed further apart, and the trailing wheels' bearings are protected from the heat of the fire. They were used for the leading wheels on James Holden's larger 2–4–0s (one of which still survives) and for the trailers on William Adam's 0–4–2s during the later 19th centuries.

American engineers meanwhile used the well-tried 4–4–0 type for most of the 19th century, aided of course by the far more generous dimensions made possible by the American loading gauge, while English designers had to put up with the narrow limits imposed by the frequent use of tunnels and overbridges constructed in the early days. At the same time North America was far ahead in the adoption of six-coupled and eight-coupled engines for heavy traffic; the first 4–6–0 appeared in 1847; an 0–8–0, admittedly of eccentric design, was built in 1835 or soon after; the first 2–8–0, the Consolidation, came out in 1865; the Pacific, or 4–6–2 type, in its initial form was introduced in 1889, and its developed form, with a wide firebox, was running by 1901.

Moreover the accommodation offered to passengers was generally much better, with the early and widespread adoption of bogie carriages, which were blocked by unreasoning prejudice both in England and on the Continent. On the other hand American tracks long remained rather primitive, with light-weight rails supported by much more frequent wooden 'ties'; and the accident record was deplorable, with the persistence of 'grade crossings', *ie* railways crossing on the level and protected by inadequate signalling. Such crossings were never common in England; the only example which persists to this day is found at Newark on the East Coast Main Line. Their prevalence in America was partly due to the multiplication of competing routes between major centres, which was often encouraged by Parliament in England, but usually with severe restrictions on the interference of one railway's traffic with another. Americans seem to be have been less inhibited!

The enginemen's comfort was also studied in the US in a way that English drivers might well envy but apparently did not, adopting a Stoic attitude towards rain and hail which is fairly well documented in the early years; until about 1860 the most that was provided was a simple windscreen, sometimes with the top bent over in a crude attempt to keep out the rain; then a short roof gradually made its appearance, and some side protection was offered by square splashers over the rear driving wheels. Typical American types giving ample protection can be seen on a Norris type 4–2–0 running in 1848, and on the very similar 4–4–0 of the same date. However the former has a normal firebox but, most unusually, outside-framed driving wheels, probably for the reason suggested above.

But the enginemen's protection was studied much earlier. Totally open footplates were used in the very early days; but a very commodious cab appears on *Pioneer*, a 2–2–2 built in 1836 for the Chicago North Western; it covers the entire rear half of the locomotive and resembles a moderate-sized greenhouse! It has two large windows on each side, each with four oblong glass panes, and three more looking out forward. The normal design also had plenty of glass and was built out, balcony-style, over the rear driving wheels giving ample width, and a roof, often in the form of a shallow gable, projecting backwards to cover the fireman at work shovelling coals.

English influence however persisted, apparently, on *De Witt Clinton*, where the footplate was entirely open but a roof was built to cover the tender. Evidently the fuel had to be kept dry, but the enginemen needed no protection!

12 – Broad gauge: Valve gears

The development of American locomotives as we have seen, followed a natural and logical progress. Leading bogies were soon adopted; the 4–4–0 type became dominant, and was freely expanded in length and weight as demands required, while the generous loading gauge allowed of much wider and higher machines. The first 4–6–0 appeared as early as 1847, though it was only towards the end of the century that ten-wheelers became common; the Pacific type, or 4–6–2, was in use much earlier than in England.

Meanwhile the British scene was far less happy. There was an enormous multiplication of different types which brought no corresponding benefits. English designers were hamstrung by the erroneous theory that a low centre of gravity was needed to avert catastrophe; and the error was compounded by the use of 'long-boiler' locomotives, which were adopted with the laudable aim of achieving economy of fuel, which indeed they achieved, but proved unstable and accident-prone as speeds increased. But before coming to these mistakes we need to describe the introduction of the broad gauge by Brunel, and the large and very efficient locomotives which it allowed, together with the reverse side of the picture, the inconvenience of having two different gauges on what should have been a single continuous railway system. This led to a bitter controversy, the so-called 'battle of the gauges', and the long-deferred but inevitable demise of the broad gauge.

The Great Western Railway was first conceived as a main line between London and Bristol; it was laid out by Brunel on exceedingly generous lines, with very easy curvature and a ruling gradient of as little as 1 in 660, except for two stretches of 1 in 100 facing up-trains. Brunel had from the first envisaged much higher speeds for passenger trains than were customary; he wanted his engines to run fast without unduly high piston speeds, and thus required fairly large driving wheels, and relatively large engines which would run stably at any speed required. With this in view he adopted a gauge of seven feet, and also designed a form of track that was guaranteed to give the necessary stability, with bridge rails carried on continuous longitudinal timbers. This soon proved too rigid, and led to broken springs; but the trouble was cured by removing the piles which were at first used to support the timbers, and by using redesigned and more flexible springs. This so-called 'baulk road' outlived the end of the broad gauge in 1892, lasting until its time for renewal.

The 20 engines first delivered to the railway were designed to meet Brunel's requirement of low piston speeds at fast travel, and were an ill-assorted lot. By far the best were two large examples of the *Patentee* type that had been ordered for the New Orleans: Railway in the USA of 5ft 6in gauge. They had boilers 8ft 6in x 4ft with some 711sq ft total heating surface, cylinders 16in x 16in and 7ft driving wheels on one, *North Star*, and 6ft 6in on the other. The Great Western engineer Daniel Gooch took these as models for his own designs from 1839 on, with cylinders 15in x 18in (later enlarged to 16in x 20in). *Ixion*, which represented the broad gauge at the trials of 1845, was of this type, with 7ft drivers.

Fig 46 – Robert Stephenson (1803–59)

Meanwhile in 1842 Robert Stephenson had begun to investigate the corrosion of chimneys and smoke-boxes, and made some experiments with the North Midland engines at Derby to discover the temperature of the smoke-box gases and of the boxes themselves. He put bits of various metals into iron cups and suspended them in the smoke-box; tin was tried, then lead and finally zinc; even this was quickly vapourised, indicating a temperature of at least 773°F, 411°C. Obviously a great deal of heat was being wasted; equally clearly, this was no new development; it had occurred in the single-flue boilers of the *Locomotion* class, where the base of the chimney often became red-hot; but locomotive designers had failed to correct it. Robert Stephenson very naturally concluded that the damage could be avoided, and massive economies achieved, if the length of the boiler tubes were increased from the normal 8–9ft to 12 or 13ft.

Long boilers were controversial from the first. Daniel Gooch had quite rightly stated that the size of the firebox was the most important factor in securing plenty of steam, but concluded, absurdly, that this was the only thing that mattered. Robert Stephenson accepted the first point. But the combination of a large firebox with very short tubes would be disastrous, as

Stephenson had shown. Nevertheless the practice of simply adding together the firebox heating surface and that of the tubes is misleading; the size of the firebox is far more important. What then was the proper length of the tubes? Modern experience has shown a length of about 14ft is on the whole the best; very long tubes, say of 20ft, yield no advantage; indeed the extra length interferes with the free flow of the hot gases, and the front end of the tubes, being relatively cool, adds little to the heating effect. But the criticism of Stephenson's tubes was carried much too far; it was even claimed that they would condense at the front end what had evaporated at the back! And there would be no appreciable drag on the flow of hot gases if fairly large tubes, say of 2in diameter, were employed. Stephenson's 12–13ft tubes were blameless in themselves; but he made unfortunate errors in design in order to secure them.

His first solution was to obtain the extra length he needed by bringing the trailing wheels forward, so that the firebox with its downward-projecting grate was behind them. The desired length of boiler could thus be achieved without any increase in the total wheelbase, which Stephenson initially seemed anxious to avoid. Again, if the boiler were lengthened, its diameter could be reduced; this would lower the centre of gravity of the engine, a state desired by almost all engineers of the time.

They seem to have feared that a higher centre of gravity would cause their locomotives to overbalance. But a derailment might be caused by several different factors, either singly or in combination. It might be due to the wheels overriding or bursting the outer rails of a curve; and once a locomotive was off the road, it would no doubt overturn, by running on to soft ground or off an embankment. It might result from a subsidence of one rail or both, or from the lurching movements of the engine caused by imperfect balancing of the machinery. There is, we shall see, ample evidence that engines deliberately built with long, slender low-pitched boilers were more liable to derail than those of the normal *Patentee* type. Nevertheless only a handful of designers were willing to risk a boiler pitched higher than usual. One of them was John Gray, who built engines for the Hull & Selby Railway in 1840 and for the London & Brighton in 1846 with boilers pitched at 6ft 2in and 6ft 3½in centres, and much later J E McConnell of the North-Western.

In some cases, indeed, a low centre of gravity actually increases the risk of derailment. This is because the outward thrust assumed to come from the centre of gravity bears more directly on the restraining rail, while only a small part of it tends to make the engine topple over. With a higher centre of gravity, within reason of course, the near-horizontal force is reduced because some of the energy is diverted into an attempt to overbalance the engine, which may in fact be a lesser risk.

In the entire course of railway history there is only one example known to me of an engine actually doing what the majority of narrow gauge designers seem to have feared, namely the disastrous accident at Salisbury on 30 June 1906, when a boat train from Plymouth approached the tight reverse curve through the station at a grossly excessive speed, certainly over 60mph. In this case the engine, of Drummond's L12 class 4–4–0, actually overbalanced without derailing; the right-hand wheels held the rails until the engine tilted over and struck a goods engine on an adjoining road. In this extreme case, the higher centre of gravity was the deciding factor; an engine of the T9 class, identical except for its smaller boiler, had escaped disaster at much the same speed; but in both cases unforegiveable risks were taken. Of course Broad Gauge engines were not liable to such a mishap, though they could derail for other reasons.

However, before proceeding to the so-called 'battle of the gauges', we need to consider the important developments in the design of valve gears, which culminated in the years 1839–41, and which follows naturally from the discussion in Chapter 5. As we have indicated, for many years the point of cut-off of the live steam entering the cylinders was fixed, and had to be settled when the engine was designed. However, the obvious need which all designers had to face was that of reversing the engine, which of course had to be done at rest. *Planet* employed loose eccentrics, as sometimes used on model locomotives; once pushed off in either direction, the two eccentrics will assume their appropriate positions. On full-sized locomotives such as the *Planets* it was necessary to disconnect the eccentrics to allow the valves to be set by hand in their proper positions. But from 1835 Robert Stephenson made use of 'gabs', *ie* V-shaped clasps projecting from the eccentric rods; as these were raised or lowered into position, a pin on the valve-rod bearing on the V-shaped sides forced the valves to assume the required position, where a parallel-sided notch in the gabs held them securely. The commoner plan was to use two eccentrics for each cylinder, giving either forward or reverse motion when engaged. The later arrangement, adopted in 1841, was to use an X-shaped member in which the two gabs were combined back-to-back; the ends of each pair of eccentric rods were connected by a link at a suitable distance, and a system of lifting rods enabled the upper joint to be lowered, and the lower one raised, into its secure position so as to connect the slide valve with the appropriate eccentric. This gear of course required a rigid valve rod in place of the jointed rods used in earlier gears.

The revolutionary new development now commonly called 'the Stephenson valve gear' dates from the next year, 1842. The basic idea was to connect the ends of the eccentric rods by a slotted link, which could be raised or lowered with the rods themselves, so that the valve rod could engage with

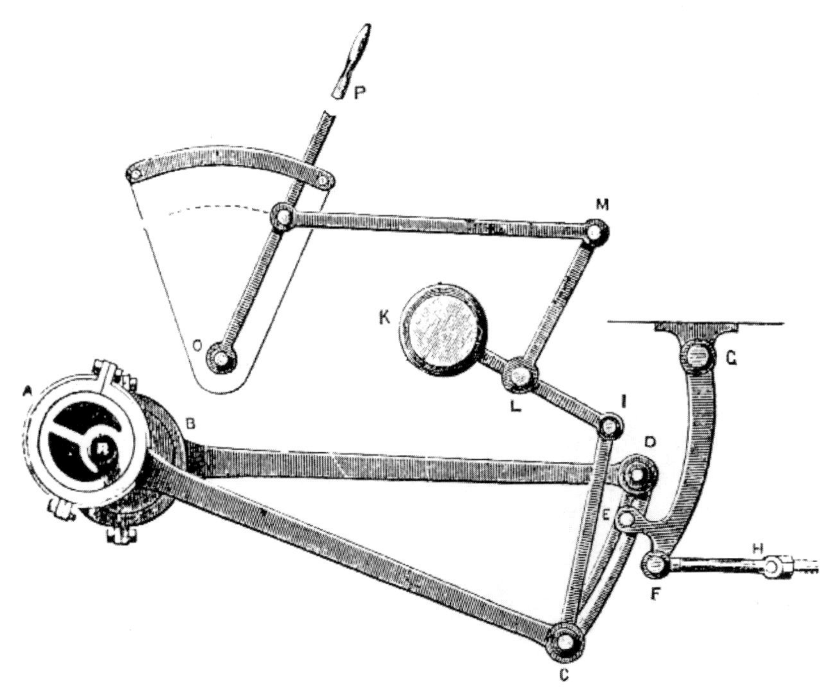

Fig 47 – Howe–Stephenson Link Motion employing a fixed pivot at G

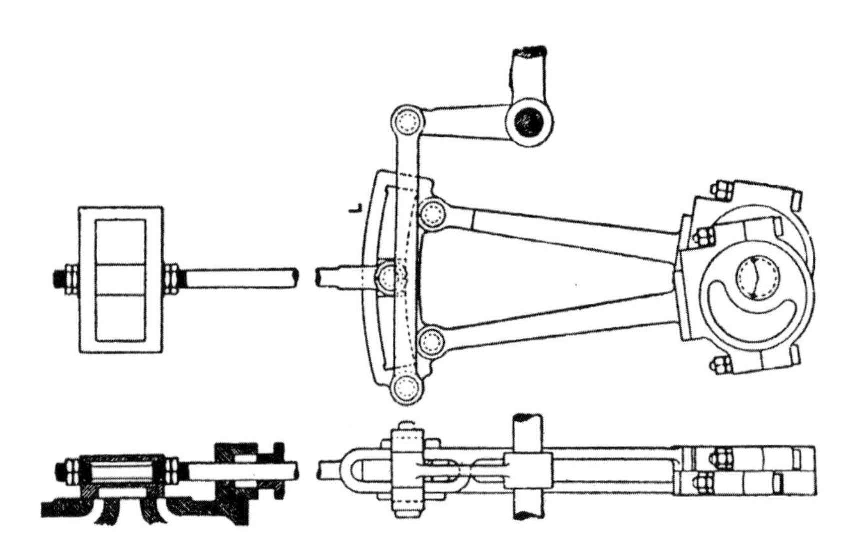

Fig 48– Howe–Stephenson Link Motion – a simpler design, supported by the pivot at the top

the top end of the link or the bottom or at any intermediate point, so as to give the valve a long or a short travel in either direction, while by bringing the mid-point of the link to the valve rod, the valves remained motionless whatever the speed of the engine.

The link itself had to be curved to fit the radius of the eccentric rods, so that they could be raised or lowered without causing the valves to move. The idea of a slotted link was apparently suggested by William Williams, while the detailed design was worked out by William Howe, and the gear manufactured at Robert Stephenson's works. The device is thus often called the Howe–Stephenson gear, but it ought properly to be renamed so that Williams's contribution is remembered. This form of valve gear remained in use till the end of steam locomotive construction; as explained above, the varying lead was helpful in meeting requirements for extra power; the Walschaerts valve gear, with its constant lead, had the advantage that it could be conveniently placed outside, and so made more accessible. An alternative to the Howe–Stephenson gear was the straight-link motion invented by Alexander Allan, which required a jointed valve-rod and a link arranged to counterbalance its up-and-down movement; a more complex arrangement, but one which made the link much easier to manufacture. Though never quite so popular, it was widely used and had a long life; engines equipped with it remained in use well into the 20th century.

13 – The Battle of the Gauges

Despite some serious faults both of character and judgement, George Stephenson possessed the priceless gift of prophetic vision. From the very beginning he had insisted on uniformity of gauge; 'make 'em all the same gauge; depend upon it, they will all be joined up one day'. Brunel, as we have seen, dissented; he was intent on making a really magnificent line between London and Bristol; he overlooked the inevitable disadvantage of the breaks of gauge which were bound to come and, with other eminent engineers, derided Stephenson's standard gauge as fit only for a 'coal-cart railway'.

With hindsight we can see that whatever the merits of Brunel's seven-foot gauge, it was already too late to propose it, as Brunel did, in 1835, when his project for the Great Western Railway was being considered. The standard gauge had come to stay by 1830, when the Liverpool & Manchester Railway was opened; it was a foregone conclusion that its extensions from Manchester to Birmingham and thence to London should be laid to the same gauge. From 1838 on, the preponderance of standard-gauge lines steadily increased; the most that the Great Western could hope for was a monopoly of the lines radiating westwards from London; from the very beginning it was hemmed in by the London & Birmingham on the north-west and the London & Southampton to the south. Contact between the two gauges was made at Gloucester in 1845, and the inconveniences of transshipment both of goods and passengers were at once exploited by the 'narrow-gauge' party, and were shown up in an exaggerated form by a well-known cartoon (see Fig 49).

Obviously a decision was needed at once, and a Royal Commission was appointed in 1845. The questions considered are summarised as follows:

(1) Whether break of gauge could be considered such an inconvenience as to call for preventive legislation.

(2) Whether such evils as might result from a break of gauge could be obviated or mitigated by mechanical means.

(3) Whether failure to devise such means would make it desirable to establish a uniform gauge throughout the country.

But in practice the problem was not always so clearly seen. A good deal of the evidence presented dealt with two other questions, which were less relevant to the immediate problem, namely:

(4) Whether a gauge of 4ft 8½in was in itself the best choice.

(5) Whether engines built to this gauge could equal the performance already achieved by broad-gauge engines.

Much of the evidence which most concerns us here relates to (5); but it was worth giving some attention to (4).

Perhaps the wisest opinion expressed was that of Joseph Locke, the engineer of the Great Western's immediate neighbours mentioned above: 'If he had to begin afresh he would adopt a gauge rather wider than the narrow gauge, but certainly not so wide as 7 feet'. The first six words were exactly to the point. But quite a number of eminent engineers, without going so far as Brunel, expressed a preference for a rather wider gauge, either in theory or in their own constructions. John Braithwaite laid out the Eastern Counties Railway to a gauge of 5ft. In the abstract this was quite a good idea. Where inside-cylinder engines were commonly used, a few extra inches between the frames would have allowed for longer axle bearings and a less constricted layout; but a non-standard gauge was too high a price to pay for such advantages.

Fig 49 – Chaos at Gloucester where the two gauges met

Other engineers went further, though without quite agreeing with Brunel. Edward Bury would have preferred another six or eight inches. John Gray, locomotive superintendent of the Brighton Railway, would have wished for a minimum gauge of 5½ to 6 feet. I believe myself that all such proposals were misguided. The standard gauge of 4 ft 8½in. may not have been ideal in theory; but any considerable enlargement would have involved serious difficulties and much needless expense. Subsequent experience, particularly in America, showed that the standard gauge could do all that was required in the way of enormously large and powerful locomotives provided the *loading* gauge, which governed their width and height, were generous enough. On the other hand, any expansion of the rail gauge would have required extra land and more massive structures in the way of bridges and tunnels even for small engines and trains, an expense which would be crippling when the railway system came to include secondary lines which could never carry heavy and lucrative traffic.

In the end, the Gauge Commission let the Great Western off rather lightly. It was not compelled to convert immediately to the standard gauge, or even to a mixed gauge; indeed the broad gauge network was allowed to expand. Broad gauge lines were laid down to Birmingham, to South Wales and to the far West; the West Cornwall Railway was actually converted from the narrow gauge so as to maintain its connection with the main system formed by the Great Western, the Bristol & Exeter, the South Devon and the Cornwall railway from Plymouth to Truro. Broad gauge persisted on some outposts even after the mixed gauge was introduced, and was not finally abolished until May 1892, when over 400 miles from Exeter westwards was converted in a single heroic operation lasting but 30 hours.

Our main interest here is of course the comparison of broad-gauge with narrow-gauge locomotives. For the narrow-gauge party, it was unfortunate that the gauge controversy broke out, as observed, in 1845. At this time the broad-gauge party could rely on the very efficient locomotives developed by Daniel Gooch from 1839 on; *Ixion*, built in 1845, belonged to this class, and was their main representative. The narrow-gauge lines had a considerable variety of locomotives in service; but the prestige of Robert Stephenson was such that his designs attracted much the most attention. As we have seen, he was at this time experimenting with long-boiler locomotives; he aimed, quite successfully, at saving fuel, but at first was unable to produce a long-boiler design that would run stably at speeds regularly achieved on the broad gauge. And it suited the broad-gauge party to keep the Commissioners' attention fixed on the performance of high-speed passenger engines; we hear surprisingly little about the performance of goods engines, where the long-boiler design had an obvious advantage; many goods trains made comparatively short runs, with frequent pauses to clear the line for fast traffic; and here the small firebox required by the long-boiler design was an obvious economy.

To consider the designs in detail, *Ixion* was a 2–2–2 of the normal *Patentee* type, with cylinders 15in x 18in, seven-foot driving wheels and a wheel-base of 12ft 4in almost equally divided, at 6ft 1in + 6ft 3in. The boiler measured 8ft 6in x 4ft with its centre pitched at 6ft 7in; grate area was 13.4sq ft, heating surface was 97sq ft from the firebox and 602sq ft from the tubes, total 699sq ft. Its main rival, Stephenson's *Great A*, had cylinders 16in x 24in, driving wheels 6ft 7in and a wheelbase of 6ft 7in + 5ft 5in, total 12ft. The boiler measured 13ft 6in x 3ft 6in, pitched at 5ft 9in; grate area was 9ft

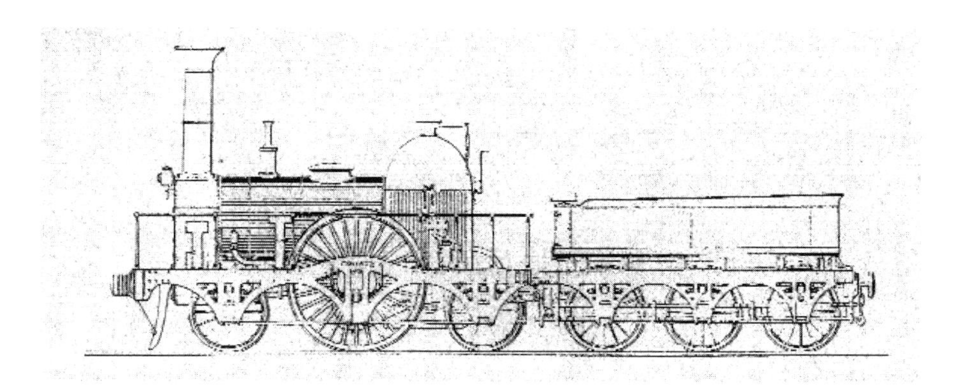

Fig 50 – Centaur *similar to* Ixion

6in, giving a heating surface of 59sq ft from the firebox and 880 from the tubes, total 939sq ft.

Robert Stephenson's long-boiler engines had been introduced in 1841, with both 2–2–2 and 2–4–0 designs, while 0–6–0 goods engines followed in 1843, all these using inside cylinders. Engines built for the Northern & Eastern Railway in 1841, of 2–2–2-type, had inside frames and a wheelbase of only 10ft 9in (6ft 4½in + 4ft 4½in), with the trailing wheels crammed in behind the drivers, so as to leave a restricted space for the firebox and footplate. An outside-cylinder variant having much the same proportions was introduced in 1843, with the object of avoiding a crank axle, but soon proved unstable at speed. *The White Horse of Kent,* a 2–2–2 of this type, had an unenviable reputation for running off the road, though its defenders claimed that it was in a run-down condition at the time of the trials. But the inside-

Fig 51 – Stephenson long-boiler locomotive

cylinder version was also prone to instability, and Stephenson seems to have recognised that the very short wheelbase was at least partly responsible; he decided that in future the minimum should be twelve feet.

This brings us back to *Great A*, as it was popularly known, though the engine's nameplate read simply *A*; the phrase meant exactly what we call a 'capital A'. It was built in 1843, and was an outside cylinder 4–2–0 with the usual rigid wheelbase; the cylinders naturally had to be moved back to a position between the two pairs of carrying wheels. The details already given show that in point of size it compared very closely with *Ixion*, though its proportions were so different; Gooch was already working on a much larger engine of the *Patentee* type, but this was not ready till 1846, too late to be considered at the trials. It appears that *Ixion* had the advantage in haulage power; with a load of 80 tons it travelled at 47mph, as against its rival's $43\frac{1}{4}$mph. But in economy of fuel *A* was the better; the evaporation of water per pound of coke was 8.8 gallons as against *Ixion's* 7.0. Robert Stephenson had no reason to be displeased with the result; Gooch's engines were speedy and reliable but economy was not their strong point.

This brings us to the vital question: whether a long-boiler engine of the *A* type, if in good condition, was adequately safe at speed. Gooch's opinion was naturally adverse; he reported to the GWR directors: 'At this speed' [47.7mph] I find this engine exceedingly unsteady, so much so that I doubted the safety of it and even deterred Mr Brunel from returning upon it, which he had proposed to do.' The Commissioners were not entirely agreed: 'The engine was less steady than that of the Great Western Railway; but the unsteadiness was unimportant. It was not related to the strokes of the piston, but seemed rather to be produced by faults in the road.'

On the other hand a remarkable run was performed in May 1847 from London to Birmingham, 112 miles in $2\frac{1}{2}$ hours, or 2 hours excluding stops, *ie* 56mph. The engine from London to Wolverton was 'one of Mr Stephenson's ordinary patent engines, and the latter part of its journey, 21 miles, was performed in 21 minutes', a remarkable feat considering the time taken to slow down with the primitive braking system then in use, which amply justifies the claim that 'the maximum speed over upwards of a mile was 75 miles per hour'. A three-cylinder engine continued the run, and did very well, running from Wolverton to Coventry, 41 miles, in 42 minutes. The load was a light one, of five carriages; but this of course does not affect the engines' claim to run steadily at high speed.

The 'Battle of the Gauges' was an important landmark; but of course there was a continuous improvement in locomotive design, especially for express passenger service. We have already noted that Gooch was planning a much larger *Patentee*, but this did not appear until 1846. For the narrow gauge, Stephenson improved his *A* type engine by adding a pair of trailing wheels. But he was prepared to desert the long-boiler type for a large *Patentee*, built in 1848 for the York, Newcastle & Berwick Railway. The leading dimensions of these three engines can be shown by a table:

It will be seen that the Stephenson 4–2–2 was a great advance on the *A*,

	GWR 2–2–2, 1846	LNWR 4–2–2, 1847	YNB 2–2–2, 1848
Cylinders	18(17)in x 24in	18in x 21in	16in x 20in
Boiler	4ft 2in x 12ft	4ft 2in x 13ft	3ft 10in x 11ft
Grate area	22.6sq ft	16sq ft	13.9sq ft
Heating surface	1952sq ft	1343sq ft	1064sq ft
Driving wheels	8ft	7ft	6ft 7in
Wheelbase	16ft (8ft+8ft)	15ft 3in (6ft 5in + 3ft 6in + 5ft 4in)	14ft 6in (7ft 6in + 7ft)

particularly in its heating surface, but was clearly outclassed by Gooch's engine, which was also much superior to the narrow-gauge *Patentee*. Thus at the Trials the broad-gauge and narrow-gauge entries were fairly equal in size; but only a year later the broad gauge had established a commanding lead; and this was further increased when the Great Western type of 2–2–2, which had too much weight on the leading wheels, was superseded by the world-famous 4–2–2s, which with little modification were in service until the end of the broad gauge.

But the long-boiler engine, in its original form with a small firebox and large overhang at the rear, was by no means obsolete. *The Edward Pease*, a 2–4–0 of this type, was built for passenger service on the S&DR in 1856, and some features of the design are strikingly recalled by the 2–4–2 tanks built for the Great Eastern Railway in 1864. For goods duties, where we have already noted the advantages of a small firebox and large boiler, the long-boiler design lasted very much longer. A long-boiler 0–6–0 constructed for the S&DR in 1874 was noted working at Whitby in 1923; it was restored to its original condition for the Railway Centenary Exhibition in 1925 and has been preserved at the York Railway Museum.

On the continent the type was perpetuated until the end of steam haulage in Austria and Spain. Even for passenger duties it was slow to disappear. A Stephenson long-boiler 4–2–0 was built for the West Flanders Railway in 1846, and was still at work in 1908.

Meanwhile a new star had been rising in the engineering firmament. T R Crampton was described by Gooch as 'a clever fellow', and was his chief draughtsman in 1837. Crampton was an ardent believer in long boilers, and in a low centre of gravity, and he had no objection to a long wheelbase.

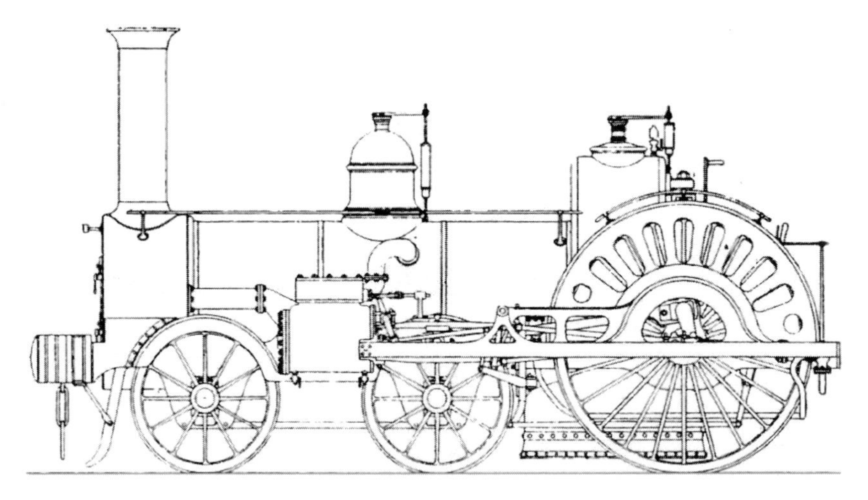

Fig 52 – Crampton locomotive Courier *(1847)*

He adopted the original plan of putting the single driving wheel behind the firebox, where its axle, contained in a casing, passed across the footplate a little below the firedoor. Thus the boiler could be lowered to a position where its bottom plates were actually a little below this axle, and need only clear the axles of the smaller rigid carrying wheels. It naturally projected a long way in front of the driving wheels, so the Crampton engines were designed not only as 4–2–0s but even as 6–2–0s if their size and weight required it.

A Crampton 4–2–0 was built for the Namur & Liège Railway in 1846. It had 16in x 20in cylinders and a total heating surface of 989sq ft, with a grate area of 14½sq ft, and 7ft driving wheels. Before delivery it was tested on the Grand Junction Railway, with the result that the newly-formed LNWR decided to use the design. *Courier* was built at Crewe works under the direction of Alexander Allan, with similar dimensions but several differences in detail, notably a prominent raised firebox intended to provide extra steam space, and Gooch's link motion. The boiler centre was only 4ft 7in above the rail level.

A larger Crampton engine, *London*, was built at Wolverton for the Southern Division of the LNWR with 18in x 20in cylinders, grate area 16sq ft, total heating surface 1529sq ft, 8ft driving wheels and 100lb pressure. By 1848 Crampton had incorporated several improvements, with double frames and the boiler supported by brackets which allowed for expansion and extra long bearings. The culmination of his designs was the very large 6–2–0 *Liverpool*, which was intended as a direct challenge to the broad gauge; it was built in 1848 and was shown at the 1851 Exhibition. It had cylinders 18in x 24in, grate area 21.5sq ft, tubes 2236, total 2290, pressure 120lbs, and 8ft driving wheels. The rigid wheelbase was 18ft 6in, and the weight about 35 tons. It hauled heavy trains with considerable success, but with its long wheelbase well loaded at each end it was very hard on the permanent way, and the design was not repeated.

Rightly or wrongly the Crampton engines were not a great success in England; in all, there were about twenty-five outside-cylinder machines of this type. It was quite otherwise in France, where the rear-driving Crampton had a great vogue; so much so that 'prendre le Crampton' became a common expression for 'catching a train'!

14 – Records and regular running

The great advantage of railways in their earliest years lay in the haulage of heavy loads. Using horses, even the primitive plateway was far better than the roads as they then were. Canals allowed even a single horse to haul a load of 50 tons; but the terrain of Great Britain was too hilly for them to be widely extended, and the frequent locks made for delays, often aggravated by winter frosts and summer droughts. Nevertheless in the early nineteenth century they had an established position, and many of the earliest railways were planned as adjuncts to existing canals; others gave access from the collieries to navigable rivers. But the ultimate success of the railway was already predictable when the route from the Bishop Auckland coalfield to the Tees was decided, and the Stockton & Darlington Railway took shape.

Locomotives, we have seen, were first used by Richard Trevithick in 1803; but it was another ten years before their usefulness was proved. In the early years, the most important consideration was the cost of working, including repairs; a load of 30 tons was acceptable, and speeds rarely rose above a smart walking pace. Stephenson was doing rather better at Killingworth by 1825; with a load of 48 tons his engine averaged 7mph on a nearly level road, with a maximum of $9^1/_2$ on the slight down-grade of 1 in 840. But it was the opening celebration of the Stockton & Darlington Railway that provided the most impressive demonstration; with a heavily overloaded train of some 90 tons *Locomotion* covered the initial stretch on slightly favouring grades at 9mph, and the up-and down continuation to Shildon at 4mph, possibly including a stop.

At Rainhill, where a regular passenger service was already foreseen, the three main competitors could all do about 30 mph running light; *Rocket* in fact concluded her tests with the prescribed load at 29mph, and drew a load of 18 tons up a 1 in 96 grade at 8mph, suggesting an average of some 20mph on the level with such a load. Some later figures are given by Pambour for the L&M *Patentees*; *Vesta*, with a load of 33 tons, averaged 27.2mph over the whole journey; *Leeds*, with over 88 tons, achieved 18.6mph, and the *Atlas* 0–4–2, with no less than 195 tons, achieved a very respectable 9.72mph. The weights include those of the tenders, but in the last two cases at least this is hardly significant. Wishaw, writing in 1839, mentions a maximum of 50mph attained by a 2–2–2 engine with a load of $22^1/_3$ tons; the average speed over $14^1/_2$ miles, discounting one stop, was $30^1/_2$mph. 50mph was also attained on the Southampton line with 46 tons on the 1 in 388 down grade near Woking. On the GWR between London and Maidenhead Stephenson's *North Star* attained a maximum of 46.8mph on the level with 30 tons load.

It was of course the Gauge Controversy which drew wide public interest to the maximum performance of engines of both types. *Ixion*, with 81 tons, ran from London to Didcot at an average speed of 47.5mph in the down direction, and 50mph on the very slight favouring grade in the reverse direction; with 61 tons the figures were 52.4mph down, and 54.6mph up with 71 tons. The maximum speeds were 60–62mph. On the standard gauge between York and Darlington, a long-boiler engine averaged about 47mph with 50 tons and 43¼mph with 80 tons. Fairly clearly the broad gauge had the advantage.

But very much better runs were recorded quite soon afterwards. In June 1846 *Great Western*, larger and more powerful than *Ixion*, ran from Paddington to Swindon, 77¼ miles, in 78 minutes with 100 tons, and the same month worked through to Exeter and back, 194 miles, taking 208 minutes running time in the down direction and 211 up. In the next year some quite interesting runs were made on the LNWR. The *Newcastle Chronicle* describes 'an extraordinary run' made on 7 May 1847: 'A special train consisting of five carriages was taken from London to Birmingham, 112 miles, in two hours thirty minutes. The actual time travelling did not exceed two hours, being an average speed of 56 miles per hour, the train being stoppped four times besides stoppage at Wolverton to change engines. The engine which started from London, No. 157, is one of Mr. Stephenson's ordinary patent engines, and the latter part of the journey, 21 miles, was performed in 21 minutes. The maximum speed over upwards of a mile was 75 miles per hour.

'The engine from Wolverton to Birmingham was also a patent engine of Mr Stephenson's having three cylinders, and it performed the first part of the journey, forty-one miles in forty-two minutes. Maximum speed in this portion of the journey sixty-four miles per hour. A side wind was blowing throughout the journey; Mr McConnel described the motion at the highest velocity as perfectly steady.'

The last sentence perhaps vindicates the LNWR track more than it does the running of Stephenson's long-boiler engines, whose overall record still seems rather patchy. Of the two stages, the second is clearly the better; the long downhill after the restart from Tring largely accounts for the high speed of the first, whereas the three-cylinder engine had to cope with the ascents to Roade summit and to the Kilsby tunnel. No doubt it was also better balanced.

However, all previous records were eclipsed by the GWR in 1848. 'From December 1847 until March 1852 the 9.50am *Flying Dutchman* was timed to run the 53 miles to Didcot at an average of nearly 58 miles per hour.

2 GREAT WESTERN.—Open to Exeter.

Miles	Down Trains.	4½	7½	6	11	8	9	10½	11	12	1½	2	4	5	5½	6½	7½	mail. 8 55	GOODS. 4½	9½	
	Trains leave	a.m	a.m.	a.m.	a.m.	a.m.	a.m.	a.m.	a.m.	noon	p.m.	p.m.	p.m.	p.m.	p.m.	p.m.	p.m.	p.m.	a.m.	p.m.	..
	PADDINGTON	..	Mail to Exeter.	6 0	..	8 0	9 0	10*15	11 0	12 0	1 30	2*0	4 0	5 0	5 30	6 30	7 30	8*55	4 30	9 30	..
6½	Ealing	..		..	..	..	9 11	Mail to Bristol.	11 11	..	1 41	..	4 13	..	5 41	6 41	7 43	..	..	..	..
7¾	Hanwell	..		..	..	..	9 16		11 15	..	1 45	..	4 19	..	5 46	6 45	7 49	..	..	..	..
9	Southall	..		W	..	..	9 20		11 20	..	1 50	..	4 24	..	5 50	6 50	7 54	..	W	..	..
13	West Drayton	..		..	..	S	9 29		11 29	..	1 59	..	4 33	..	6 0	6 59	8 3	9 20	5 24	..	..
18	SLOUGH	..		6 35	..	8 25	9 45	10 53	11 46	12 38	2 15	2 38	4 46	5 33	6 12	7 15	8 16	9 30	5 44	10 30	..
22½	MAIDENHEAD	..		6 45	..	8 48	9 56	11 3	11 55	..	..	2 50	4 56	..	6 30	..	8 26	9 42	6 0	..	..
31	TWYFORD	..		..	..	9 5	..	..	..	1 8	..	3 8	5 12	6 3	..	..	8 42	..	..	11 14	..
36	READING	..		7 10	..	9 18	..	11 35	..	1 20	..	3 22	5 25	6 17	..	..	8 56	10 10	7 20	11 40	..
41½	PANGBOURNE	..	..	..	..	9 22	..	..	..	1 37	..	S	..	6 30	..	..	9 10	..	7 40	..	..
44½	Goring	..	..	7 28	..	..	..	..	..	1 47	..	..	..	S	..	..	9 18	..	..	..	..
47¾	WALLINGFORD ROAD	..	..	..	..	9 46	..	11 58	..	..	..	3 45	..	6 45	..	..	9 27	10 35	..	12 30	..
53	DIDCOT (Junction)	..	..	7 44	..	9 58	11 15	12 12	..	2 6	..	3 58	..	6 58	..	..	9 40	10 48	8 20	12 45	..
55	APPLEFORD	..	..	..	..	10 10	..	..	..	2 14	..	..	..	..	..	..	9 50	..	10 10	..	..
56	ABINGDON ROAD	..	..	7 55	..	10 13	..	12 23	..	2 17	..	4 10	..	7 10	..	..	9 55	11 0	10 13	..	..
63	OXFORD	..	..	8 10	..	10 30	11 36	12 38	..	2 35	..	4 25	..	7 25	..	..	10 10	11 16	10 30	..	..
56½	STEVENTON	..	..	7 50	..	..	..	..	..	2 17	..	..	..	7 8	..	..	..	..	..	..	..
63½	FARRINGDON ROAD	..	..	8 7	..	..	..	12 58	..	2 32	..	4 23	..	7 23	..	..	..	11 12	..	1 19	..
71½	SHRIVENHAM	..	..	..	..	10 40	..	..	..	2 50	..	4 41	..	7 40	..	..	..	..	9 30	..	..
77	SWINDON (Junction)	..	..	8 40	..	10 55	..	1 6	..	3 6	..	4 55	..	7 50	..	..	..	11 40	10 0	2 16	..
	CHELTENHAM (Departs for)	..	..	9 2	..	11 7	..	1 17	..	3 17	..	5 7	..	8 2	..	..	..	11 52	11 7	9 2	..
81½	Purton	..	..	9 10	..	..	..	1 27	..	..	..	5 17	..	..	..	..	..	..	..	9 10	..
85½	MINETY	..	..	9 20	..	11 23	..	1 38	..	3 32	..	5 27	..	8 22	..	..	..	..	11 23	9 20	..
96	CIRENCESTER	..	..	9 50	..	11 50	..	2 5	..	4 6	..	5 53	..	8 50	..	..	..	12 35	11 50	9 50	..
—	SWINDON Junction (Departs)	Passengers and Goods	..	8 50	..	11 5	..	1 15	..	3 15	..	5 5	..	8 0	..	..	..	11 50	11 10	2 30	..
82¾	WOOTTON BASSETT		..	9 3	..	..	..	..	..	3 27	..	5 18	..	8 12	..	..	..	..	11 40	..	..
93¾	CHIPPENHAM		..	9 28	..	11 35	..	1 48	..	3 50	..	5 40	..	8 33	..	..	..	12 20	12 30	2 43	..
98¼	Corsham		..	9 38	..	11 47	..	..	..	4 2	..	..	..	8 48	..	..	..	..	..	..	..
101¼	Box		..	9 48	a.m.	11 57	p.m.	..	..	..	p.m.	6 0	..	8 52	..	p.m.	..	..	..	..	..
106¾	BATH	4 30	*7 15	10 0	11 0	12 10	1 0	2 10	..	4 20	5 30	6 15	..	9 10	..	9*30	..	12 50	1 10	4 20	..
108¾	Twerton	..	7 20	..	11 5	..	1 5	..	..	..	5 35	..	..	..	..	9 35	..	..	..	..	..
111¾	Saltford	..	7 28	..	11 13	..	1 12	..	..	..	5 45	..	..	..	..	9 42	..	..	..	..	..
113¾	Keynsham	..	7 35	10 20	11 20	..	1 20	..	..	4 35	5 50	..	..	..	..	9 49	..	..	..	..	..
118¼	BRISTOL arrival	5 20	7 45	10 30	11 30	12 35	1 30	2 30	..	4 50	6 0	6 40	..	9 40	..	10 0	..	1 15	1 45	6 20	..
	BRISTOL departure	..	8 0	10 50	..	12 40	..	2 45	..	5 5	..	9 55	..	..	..	..	..	1 25	..	a.m. 7 0	..
126¼	Nailsea	..	8 18	..	..	1 8	..	..	..	5 23	..	7 13	..	..	..	..	..	..	..	..	..
130¼	CLEVEDON ROAD, Yatton Junc.	..	8 28	11 18	..	1 18	..	3 8	..	5 34	..	7 23	..	..	..	..	..	..	..	7 45	..
133¾	Banwell	..	..	11 25	..	..	..	..	..	5 43	..	..	..	..	..	..	..	..	..	7 58	..
138¼	WESTON SUPER MARE	..	8 45	11 35	..	1 28	..	3 23	..	5 50	..	7 40	..	..	..	..	..	..	..	8 10	..
146¼	HIGHBRIDGE, near Burnham	..	9 3	11 53	..	..	..	3 41	..	6 10	..	7 58	..	..	..	..	..	..	..	8 42	..
151¼	BRIDGEWATER	a.m	9 20	12 10	..	2 0	..	3 56	..	6 25	..	8 15	..	..	..	..	..	2 35	..	9 5	..
163	TAUNTON	8*0	9 40	12 30	..	2 25	..	4 15	..	6 50	..	8 35	..	..	..	..	..	3 0	..	10 10	..
170	WELLINGTON	8 15	9 55	12 45	..	..	..	4 30	..	7 5	..	8 50	..	..	..	..	..	3 15	..	10 55	..
179	TIVERTON ROAD	8 34	10 15	1 10	..	2 55	..	4 50	..	7 30	..	9 15	..	..	..	..	..	3 35	..	11 30	..
181¼	COLLUMPTON	8 39	10 20	1 17	..	3 0	..	..	..	7 37	..	9 22	..	..	..	..	..	3 40	..	11 40	..
185¼	Hele	8 49	10 28	1 27	..	3 10	..	5 3	..	7 47	..	9 32	..	..	..	..	..	..	..	12 0	..
193¾	EXETER	9 0	10 45	1 45	..	3 30	..	5 20	..	8 5	..	9 50	..	..	..	..	..	4 5	..	12 30	..

Fig 52 – GWR Timetable, 1844

On 11 May 1848 the train carried a group of scientific observers and ran to Didcot in 47½ minutes, an average of 66 mph. Top speed must have been 80mph.' It should be noted that this run was made by a regular train, not a light-weight special, and that the line is partly level, partly on slight adverse grades of 1 in 660; thus the speed of the train cannot have been constant; the maximum speed must have been 75mph, and 80 is not impossible.

From these impressive records we may turn back ten years to the much lower speeds advertised to the public and recorded in the time-tables. In 1838 the Great Western was decidedy uninformative, perhaps wisely considering the initial state of its locomotive fleet; departure times of the trains to Maidenhead are given, but no promises are made about arrivals! By 1841 matters had greatly improved; the down night mail to Bristol, 118¼ miles, took 4h 10m, an average of 26.4mph. The Birmingham line was much slower; the journey of 113 miles took 5¼ hours, averaging only 21.3mph. 'Soon after 1846' the time came down to 2½ hours to Bristol and 4h 25m to Exeter; by 1850 there was a slight easing, 2hrs 35 to Bristol (an average speed of 43.1mph) and 4h 40m to Exeter. To Birmingham now took 3h exactly, averaging 37.7mph.

We can now summarise the service offered by the railway to the most important towns, omitting Bristol and Birmingham as already treated, using the reprint of Bradshaw's guide for March 1850. My list reads:

London to Manchester	188¼	5h 40m	33.2mph
Leeds	206½	5h 35m	36.0mph
Norwich	126	4h 45m	30.0mph
Dover	88	2h 30m	35.2mph
Brighton	50¼	1h 15m	40.2mph
Southampton	80	2h 55m	27.4mph
Liverpool to Manchester	31½	55 min	34.2mph
Birmingham to Bristol	90½	4h 20m	20.8mph

In 1850 the number of competing routes was comparatively small. But examples can be found, some of which follow unfamiliar pathways:
To conclude this chapter, it may be useful to give a table showing the best speeds achieved on various lines in offering these services. In this case, it is difficult to make a fair allowance for runs with one or more stops, against non-stop runs. As a rough approximation, I suggest five minutes per stop, if only a brief call is indicated, with an extra allowance for a longer stop. It should be remembered that with the primitive braking system then in use it would take some time to bring the train to a halt. I give the booked times first, and the speeds involved, followed by the corrected times and speeds.
Obviously the broad gauge is an easy winner; though it is interesting to find the Bristol & Exeter showing up marginally better than the Great Western, if my allowances are to be trusted, despite the formidable hindrance of the climb to Wellington summit. No doubt the loads were rather lighter west of Bristol. On the narrow gauge, the Brighton line does notably well, considering the long switchback grades of 1 in 264; whereas the York, Newcastle & Berwick fails to profit by the dead-flat racing stretch between York and Darlington, where a one-hour timing should have been possible. Further north the Edinburgh & Berwick makes a better showing over a difficult route including the even more daunting climb over Cockburnspath summit. Of the three Southern main routes the Southampton line is completely out of the running, as its best express begins with a leisurely run to Woking, a level run of 28 miles in 45 minutes, 37.7mph, and stops at virtually every station thereafter; a dismal beginning if one is bound for Dorchester! This is a classic case of unenterprising management; Southampton should have been given at least one train making four stops only, Woking, Basingstoke, Winchester and Eastleigh, which with a brisk start from London could have brought it quite near the two-hour schedule proposed by Joseph Locke when the line was opened.

In compiling these tables the 1850 Bradshaw has proved an indispensible guide; but it should not be trusted in every particular. A glance at the Dover line shows two trains booked to cover the five miles from Folkestone to Dover in five minutes, average 60mph! A glance at the up-line table apparently deepens the mystery; the trains are booked to call at Folkestone at the same moment that they leave Dover! Obviously it is the Folkestone times that are at fault; the best times between Dover and Ashford are the perfectly reasonable 35 minutes non-stop and 37 with one stop for the 21 miles.

The survey has to stop somewhere; but 1850 has its disadvantages. The English railway network was about to be transformed by the opening of the Great Northern Railway; the vital stretch from London to Peterborough gave access to the Lincolnshire lines and a reasonable route to the north even without the continuation northwards from Peterborough. Again, it is interesting to contrast the rapid development of railways in Scotland with their almost complete absence from Wales, where the only routes open were the Holyhead main line in the north and a few coal-carrying lines in Glamorganshire. Even the vital route from Gloucester to Cardiff was still to come. And even when the railway network reached its fullest development there never was a convenient through route between North and South Wales; one had to enter England and travel via Shrewsbury. Had the mountains been less obstructive, a good railway connection might have done much to reduce the feeling of disunity between the two breeds of Welshmen.

15 – Conclusion

Fig 53 – Bulkeley, *a broad-gauge GWR 4–2–2 locomotive as built from 1847 onwards*

Fig 54 – Jenny Lind *(1847)*

I have fixed on 1850 as the proper date for ending this recital. It does not of course mark any pause in the development of British locomotives in size and power; they continued to increase, with the one exception of the broad gauge, where the large outside-framed 4–2–2 headed express passenger trains down to 1892. Nevertheless there was a notable change; the 1840s had been a time of great variety and alteration; from 1850 British design followed a much more consistent pattern. Yet this was quite unlike the uniformity seen in America, where the outside-cylinder 4–4–0 was virtually unchallenged for forty years in passenger service, apart from some 2–6–0s. The great majority of British designs in this period were indeed six-wheelers, but of three main types; engines with outside frames whose history has already been briefly sketched, and inside-framed machines, with a choice between inside and outside cylinders.

Passenger engines might be 2–2–2s, 2–4–0s or 0–4–2s. The pattern for 2–2–2s in the next decade was set by *Jenny Lind*, designed by David Joy and built by E B Wilson in 1847. She had cylinders 15in x 20in, and weighed a little over 24 tons. These proportions seem modest enough; but she had a boiler pitched at 6ft 5in. (Ahrons' figure is 5ft 9in; but this is clearly an error, since the drawing shows that the pitch is greater than the diameter of the 6-ft driving wheels.) The grate area, 18.9sq ft, was more than either of the large Stephenson engines noted above, and the firebox was widened so as to overlap the frames, and also slightly raised, giving a firebox heating surface of 80sq ft out of a total of 800. The great success of the *Jennies* was attributed, quite rightly, to their high boiler pressure of 120lbs. But it was the sound design of the heating surface that enabled the pressure to be sustained.

The last 2–2–2 was built in 1894, and lasted about ten years; the other two types lived much longer. We have noted the long service of the Midland outside-framed 2–4–0s; equally notable was that of the inside-framed mixed traffic 2–4–0s built between 1891 and 1902 by James Holden for the GER; robust and free-running machines which could be seen in their later years not only at the sleepy backwaters of Mildenhall and Wells-next-the-Sea, but climbing the windy heights of Stainmore summit in the Pennines. World-famous were the *Gladstone* class 0–4–2s, which could still handle very heavy express trains on the Brighton line in the 1920s. Less glamorous but widely adaptable were the mixed traffic 0–4–2s built by William Adams for the LSWR. Perhaps the best known of all 0–4–2s was *Lion*, shown in our frontispiece, an outside-framed machine built for the Liverpool & Manchester Railway in 1838, which after many years of degradation and obscurity was rescued and reconstructed, and played a star part in that lovable comedy film *The Titfield Thunderbolt* (1952). *Lion* also led the *Rocket 150* cavalcade in 1980 to commerate the opening of the L&MR in 1830. But apart from a few specially preserved locomotives, long survival was quite foreign to engineers in the States and Canada, for whom twenty years of hard service was usually enough.

The 4–4–0 type only became prominent in England in the 1870s, though examples from the 60s can easily be found. An important factor in bringing it into favour was the long-wheelbase bogie with controlled side-play invented by William Adams; this effectively killed the suspicion of early designers that the bogie might swing round broadside-on to the track and derail the engine. The totally unreasoning dislike of bogie *carriages*, where the geometrical objections did not apply, persisted much longer among British engineers who should have known better.

The British 4–4–0 was of three types. First, outside-framed machines, already noticed above; the most illustrious example was the record-breaking *City of Truro*, which has been preserved at York and is too well known to need description. Second, outside-cylinder layouts, avoided by many designers as unsteady runners; but efficient and hard-working examples were built over many years for the Highland Railway from 1873, and nearer home by William Adams for the LSWR from 1879 to 1896. The last example was running until 1945, and again was set aside for preservation and restored to its magnificent livery.

However the inside-cylinder 4–4–0 was by far the most popular, and came to be regarded as the classic example of British design, seen at an early date in *Abbotsford*, built by Dugald Drummond for the North British Railway in 1876. Its direct descendants were the Caledonian *Dunalastairs*, which included a small class built for the Belgian railways, where their shapely lines contrasted with the horrific ugliness of the native machines. A large class of 66 engines was built by Drummond for the LSWR in 1899–1901, and proved most successful. Rebuilt and superheated after the First World War they became ubiquitous on the Southern Railway, displacing many of the native products of Ashford and Brighton. The whole class was in service for fifty years, and one of them, No. 120, was kept in readiness to work the Royal Train, and once again has been preserved at York.

From this brief look at locomotive development from 1850 onwards we may turn back to the early years; our picture is still incomplete without some reference to the men whose task it was to drive and fire the engines, and if need be to effect emergency repairs. In the earliest years the railways' close association with collieries was reflected in their train crews; sometimes former coalmen themselves, but also men who had gained their experience by the management of steam-worked pumping machinery. The pitmen were used to extremes of heat and cold; the freezing walk to the pithead in the early morning, the stifling heat below ground. So they saw nothing unusual in a footplate entirely unprotected from gales of wind, rain and snow assailing their heads and shoulders, not to mention inconvenient warmth lower down. Right through the 1850s John McConnell was building engines with no vestige of a cab. But it was about 1850 that some protection was first offered in the form of a simple wind-shield; ten years or so later this was improved by bending the front plate backwards, so as to offer some overhead protection. But some drivers disliked even this; on the Midland Railway the men went to Kirtley and protested at the drumming noise made by the unsupported plate. A rudimentary cab with a roof and some side protection dates from about 1870. Even so, there was often a D-shaped cut-out made in the side plates, and many drivers preferred to sit or crouch leaning out a few inches so as to gain a direct view forwards. The side-window cab which blocked this opening was officially hailed as an improvement, but was not always popular. In this whole development British engineers, and many continental ones, were at least twenty years behind the Americans; and throughout the nineteenth century most English cabs were narrow enough to allow an engineman to creep forward along the footplate and attend to the cylinders if needed.

We have seen how the earliest engine crews were drawn from the pitmen. But the rapid expansion of the railways in the 1830s made it necessary to recruit drivers and firemen with little or nothing in the way of previous experience; no formal scheme of training was ever provided for British engine-drivers; they learnt their craft as best they could by imitating those of their mates who achieved the best results on the score of punctuality and economy of fuel; the latter in particular was closely monitored and studied by directors. And as the railways expanded, their management often passed into the hands of men who were not themselves practical engineers, but entrepreneurs interested mainly in securing quick profits. An obvious expedient was to exploit the train crews and other employees by paying minimal wages and imposing long working hours; though it must be said that the practical men, such as Daniel Gooch on the Great Western, were by no means averse to cutting the men's wages when economies were called for. And as to working hours, there seemed to be no limit to the length of a day; there are plenty of examples of drivers remaining on duty for 25 hours at a stretch. And whereas an overtired driver could often carry on and make the right movements through his intuitive knowledge of the engine, the signalman had no such resource; there are heart-rending stories of accidents brought about by men who had realised their danger and applied in vain for relief.

It was only very gradually as the nineteenth century wore on that the railway workers learnt to unite and to protect one another. Certainly there were some early examples of mutual help; thus in 1843 Daniel Gooch set up a Locomotive Department Sick Fund, and the next year a GWR Mechanics

Institute to provide books and lectures, after it was noticed that the men had already started their own library. The Enginemen and Firemen's Mutual Assurance Society was started in 1867 by a few GWR enginemen with the assistance of Daniel Gooch. But Gooch was almost a lone example of such paternalistic interest in his men's welfare; his brother John Gooch on the Eastern Counties Railway became notorious both for lining his own pockets by selling the company's property to outside firms and for his harsh treatment of his men; 'he reduced costs by sacking men and offering them their jobs back at lower rates, and by deducting from those reduced wages large fines as a punishment for anything that went wrong with their engine or train, from a lack of punctuality to the snapping of a coupling or the collapse of a connecting-rod'– so Adrian Vaughan. As a result, 178 drivers, firemen and fitters gave notice to quit unless Gooch were removed. Gooch stayed on and found other men, but also blacklisted the 178 so that they would not be employed on railways elsewhere. It was another six years before shareholders took notice and removed Gooch and his accomplice David Waddington, 'and they went on to a life of luxurious retirement'.

Gradually relations between railwaymen and managers improved, and there are many examples of the men's fiercely exclusive pride in, and loyalty to, their own companies, which to some extent persisted even after the 1923 grouping merged the smaller companies into four groups. This was particularly true of the Great Western, which was the least affected by the grouping; the very distinctive Swindon tradition of locomotive building continued unaltered, and the Western Region continued to perpetuate its elegant pattern of lower-quadrant semaphore signals long after they had officially become obsolete. On the locomotive side a fairly amicable cooperation was achieved on the Eastern and Southern groups, largely through the wisdom of Nigel Gresley and R E L Maunsell, who freely adopted the best of the pre-grouping designs, whereas LMS practice was for many years retarded and embittered by an obstinate rivalry between the partisans of Crewe and of Derby.

It has to be said that almost to the end of steam working two factors combined to obstruct British locomotive design; first, the intense rivalry and secrecy which hindered cooperation between the designers of different companies; second, the empirical approach to design, which seldom made sufficient use of basic scientific research, so far from making this accessible to enginemen; so unlike the exclusive French race of *mechaniciens*. But we cannot deny that all this contributed greatly to the abiding interest of British locomotive history, and to the study of the enginemen themselves, many of whom were acknowledged masters of their craft, and could make allowance for the idiosyncrasies of particular engines. In the end the steam locomotive was banished from the national railway network; belated research had shown that the limits of improvement had been reached; though there is no possible defence of the over-hasty process of dieselization forced upon the railways by ignorant politicians. However, within a few years, as the movement for preservation gathered momentum, even British Railways were compelled to realise the abiding public appeal of the steam locomotive, while not a few enginemen went to work on the preserved steam-worked railways, happily joining the well-instructed amateurs who had learnt the secrets of their craft.

And a fascinating craft it was, even though the preserved steam trains generally went at a gentle amble – a far cry from the hard graft of running an express service day after day, often with run-down and ill-maintained engines. But the steam-engine driver often developed a knack of coping with all but the gravest emergencies. Most students of railway practice have their favourite anecdotes of resourcefulness in coping with difficulties. I will end with two of mine.

One of them repeats the old fable of the mouse coming to the rescue of the lion. One day in the late forties the engine of the Cornish Riviera Express failed a few miles West of Reading. The engine crew commandeered a minute 2–4–0 which was standing with a local goods train in a siding; built by Beyer Peacock in the nineties for the long defunct Midland & South-Western Junction Railway, an early idol of my own. They had to begin by detaching and shunting the cripple; then release the brakes, not always an easy task; after which the little 2–4–0 drew the 500-ton train to Newbury, fortunately an easy run. There a stout-hearted 4300 class 2–6–0 was available to proceed over Savernake summit and down to Westbury, where at last an express engine was waiting, hastily summoned from Bristol. Total time lost, something over a hour; but the delay might have been much longer if a relief engine had been summoned at once

In passing, this event inspired a fine flight of fancy from Colin Maggs (*The MSWJR* p129): 'In 1927, when No 6003 *King George IV* [built 1928!] pulling the Cornish Riviera Express derailed a bogie at Theale, the ex-MSW engine took the train on to Westbury'. Obviously no one in his senses would have entrusted this 500-ton train to so small an engine for the fifty-mile run to Westbury, including the toilsome ascent to Savernake, with three miles beyond Bedwyn steepening from 1 in 175 to 1 in 106 at the summit. Her weight limit on her home ground was 130 tons! Some people will believe anything.

The other episode was one which I witnessed myself. A fast train from Carlisle to Newcastle, drawn by a three-cylinder *Hunt* class 4–4–0, came to a stand halfway up the climb to Gilsland; part of the Walschaert's gear had come adrift. and the flailing rod was threatening further damage. It took the engine crew about fifteen minutes to detach the rod and stow it in the tender, and to secure the valves of the useless cylinder in mid-gear; after which we got going on the two remaining cylinders and arrived at Newcastle only twenty-five minutes late. Here again a much longer delay had been avoided by an engineman's resource.

The study of the engine-driver's art may well help the amateur enthusiast and participant to acquire the their characteristic gifts of vigilance and resourcefulness. Vigilance is a Christian virtue, commended by St Peter in his First Letter; resourcefulness has I think been rather underestimated by Christian tradition. But these two qualities both enrich our lives and have proved again and again to be essential to our national wellbeing.

Bibliography

E L Ahrons, *The British Steam Railway Locomotive*, 1825–1925, London 1927, repr. 1987
Mark Archer, *William Hedley*, third edition, Newcastle 1885
Rodney Dale, *Early Railways*, London 1994
H W Dickinson, *A Short History of the Steam Engine*, Cambridge 1939, repr 1963
— *James Watt, Craftsman and Engineer*, Oxford 1927, repr. Cambridge 1936
— and A Titley, *Richard Trevithick Memorial Volume*, Cambridge 1933
C Hamilton Ellis, *The Pictorial Encyclopedia of Railways*, Feltham 1968
H Loxton, *Railways*, London 1963, repr 1968
O S Nock, *Railways Then and Now*, New York 1975
L T C Rolt, *George and Robert Stephenson; the Railway Revolution*, London 1960
— *The Cornish Giant; the story of Richard Trevithick*, London 1960
J B Snell, *Early Railways*, London 1964, repr 1972
— *Mechanical Engineering* (Railways), London 1971
C E Stretton, *The Development of the Locomotive*, London 1896, repr 1989
Adrian Vaughan, *Railwaymen, Politics and Money*, London 1997
J G H Warren, *A Century of Locomotive Building*, Newcastle 1923; repr Newton Abbot 1970 with introduction by W A Tuplin
N A Wood, *A Practical Treatise on Railroads*, London 1825, 1831, 1838

Index

Note: where items appear on adjacent pages,
only the number of the first is given.

Adams, William: 104
Agenoria: 48
Allan, Alexander: 56, 89, 96
Atlantic: 76
Atlas: 71, 98
Baltimore & Ohio Railway: 73, 75
battle of the gauges: 87, 90
Bernina Railway: 26
Best Friend of Charleston: 76, 78
Birmingham & Gloucester Railway: 80
Blackett, Christopher: 25, 39
Blenkinsop, John: 17, 25, 39
Blucher: 39
Bobbie: 42
Bolton & Leigh Railway: 53, 63
Booth, Henry: 54, 65
Boulton, Matthew: 11, 15
Braithwaite, John: 91
Branca, Giovanni: 3
Bulkeley: 103
Bulldogs: 82
Bury, Edward: 68, 71, 79, 91
Camden & Amboy Railroad: 73, 78
Campbell, Henry R: 79
Canterbury & Whitstable Railway: 67
Catch-me-who-can: 22
Chaloner: 76
Chittaprat: 51
Cities: 82
City of Truro: 80, 105
Coalbrookdale locomotive: 19, 32
Courier: 96
Crampton, T R: 96
Cugnot, Nicholas: 13
Davy, Sir Humphry: 43
De Witt Clinton: 76, 83

Der Adler: 70
Dick, Professor Robert: 10
Dickinson, HW: 110
Dickinson, Robert: 23, 110
Dixon, John: 59, 63
Duke of Cornwall: 80
Edinburgh Museum: 30
Edward Pease, The: 95
Experiment (coach): 51
Experiment (locomotive): 79
Flying Dutchman: 99
Gateshead locomotive: 20
Geordie: 42
Gladstone: 104
Goliath: 68
Gooch, Daniel: 85, 92, 94, 106
Gooch, John: 107
Gray, John: 86, 91
Great A: 92, 94
Great Western Railway: 80, 82, 84, 90, 94, 99, 103, 106
Gresley, Nigel: 107
Hackworth, Timothy: 12, 48, 52, 57, 61-64, 71
Hector: 71
Hedley, William: 17, 25-30, 32, 39, 48, 110
Hero of Alexandria: 1
High Peak Railway: 27
Holden, James: 82, 104
Howe, William: 88, 89
Hull & Selby Railway: 86
Hunt: 108
Ironbridge Gorge Museum: 20
Ixion: 85, 92, 99
Jenny Lind: 82, 103
Jervis, John B: 79
Jimmie: 42
John Bull: 77
Joy, David: 104
Kilmarnock & Troon Railway: 41
King George IV: 108
King Leopold: 70
L'Éléphant: 70

La Victorieuse: 71
Lancashire Witch: 54, 65, 67, 72
Laxey Wheel: 4
Leeds: 98
Leeds & Thirsk Railway: 71
Leicester & Swannington Railway: 69, 71
Leupold, Jacob: 26, 31
Lion: frontispiece, 104
Liverpool: 69, 71, 79, 97
Liverpool & Manchester Railway: 42, 50, 53, 57, 68, 72, 90, 101
Llewellyn drawing: 19, 21, 32
Locke, Joseph: 42, 91, 102
Locomotion: 12, 41, 45, 47, 85, 98
London: 96
London & Birmingham Railway: 42, 69, 73, 90
London & Brighton Railway: 86
London & Southampton Railway: 42, 90, 101
Losh, William: 40
Maunsell, R E L: 107
McConnel, J E: 86, 99, 106
Meteor: 67
Middleton Railway: 25
Midland Railway: 82, 106
Mohawk & Hudson Railroad: 76, 79
Morgan, John: 10
Murdock, William: 13, 31
Murray, Matthew: 25, 31, 39
Namur & Liège Railway: 96
Newcastle & Berwick Railway: 95, 102
Newcastle Chronicle: 99
Newcomen, Thomas: 6-8, 10, 38
Nicholas, Grand Duke: 71
North Star: 85, 98
Northumbrian: 66
Novelty: 57-61, 68
Old Ironsides: 78
Outside-framed American: 80
Patentee: 56, 69, 79, 84, 86, 92, 94, 98
Pease, Edward: 46
Penydaren engine: 19, 39
Perseverance: 57

Philadelphia & Reading Railway: 73
Phoenix: 67
Pioneer: 81, 83
Planet: 68, 75, 78, 87
Porta, Giambattista della: 4
Puffing Billy: 12, 29, 30
Rocket: 45, 53-7, 59, 63-7, 98
Royal George: 12, 48, 51, 61
Samson: 68
Sanspareil: 57, 61
Savery, Thomas: 4
Science Museum: 11, 20, 30
Séguin, Marc: 48, 54, 65, 72
slide valve: 31, 41, 48, 54, 61, 87
Smiles, Samuel: 42
Somerset, Edward: 4, 6
Stephenson long-boiler locomotive: 93
Stephenson's second locomotive: 40
Stephenson, George, and his son Robert, occur frequently throughout
Stockton & Darlington Railway: 45, 51, 72, 98
Stourbridge Lion: 48, 72
Tayleur & Co: 70
Titfield Thunderbolt, The (1952): frontispiece, 104
Tom Thumb: 75
Trevithick, Richard: 1, 6, 16, 26, 32, 39, 98
Twin Sisters: 54
Twining, E W: 41
Typical American: 81, 83
Vesta: 98
Walschaerts gear: 35
Watt, James: 2, 6, 10-13, 15, 29
Watt, Thomas: 9
Wellington: 39, 41
White Horse of Kent, The: 93
Wilberforce: 53
Williams, William: 89
Wilson, E B: 104
Wood, Nicholas: 38, 42, 46, 59
Wylam colliery: 26
Wylam Dilly: 30
Wylam plateway: 27

Acknowledgements

Special thanks are due to Rodney Dale and John Snell, both of them accredited authorities on Railway matters, for expert and timely advice.

We are particularly grateful to David Rodgers for allowing us to use his picture of *Lion* on the jacket and as the frontispiece.

Most of the other illustrations are reproduced from out-of-copyright sources; we have done our best to trace other copyright holders, and apologise if we have overlooked anyone; please contact us and we will be pleased to amend any future printings.

Finally, we thank Charlotte Edwards for her page layouts, and Chris Winch for assembling the jacket files.

A note on units

In this book, we have used the Imperial units that our pioneers of engineering would have used. We believe that metric equivalents clutter the page and detract from the history, and we make no apology for omitting them.